Site management (IYCB 2) Handbook

SITE MANAGEMENT

(IYCB 2) HANDBOOK

Claes-Axel Andersson
Derek Miles
Richard Neale
John Ward

International Labour Office   Geneva

Andersson, C. A., Miles, D., Neale, R. and Ward, J.
Site management (Improve Your Construction Business 2) Handbook
Geneva, International Labour Office, 1996
/Management development/, /Guide/, /Management/, /Construction/,
/Small scale industry/, /Construction industry/. 12.04.1
ISBN 92-2-108753-0
ISBN for complete set of two volumes: 92-2-109315-8
ISSN 1020-0584

ILO Cataloguing in Publication Data

PREFACE

The *Improve Your Business (IYB)* approach to small enterprise development has proved its worth in many different countries, and has demonstrated the need for publications which are written simply and clearly but which can still communicate the basic management knowledge required by entrepreneurs if they are to run small businesses successfully.

Although all small businesses face some common problems and certain management principles are universal, experience has shown that a sector-specific development of the IYB approach would be widely welcomed.

This demand was particularly strong from enterprises in the construction sector, since small contractors have to cope with the special managerial problems that arise from bidding for and carrying out varied and dispersed projects. They are also faced with highly cyclical demand.

The ILO has responded by developing this Improve Your Construction Business (IYCB) series to suit the specific needs of small building and public works contractors. The IYCB series of three handbooks and three workbooks is available either separately or as a set, and comprises:

1. Pricing and Bidding (IYCB 1) Handbook and Workbook
2. Site management (IYCB 2) Handbook and Workbook
3. Business management (IYCB 3) Handbook and Workbook

They have been designed for self-study, but there is also an IYCB trainer's guide to assist trainers in preparing for and running seminars and workshops. As demand emerges, further handbooks and workbooks will be added to suit the specialist needs of, for example, road contractors and materials manufacturers.

The first handbook and workbook deal with pricing and bidding to obtain new projects. Too many contractors produce "guesstimates" not estimates of project costs, so they either bid too high and lose the contract or – often even worse – get the work at a price which is below cost. The first handbook takes the reader step-by-step through the preparation of the bid for a small building contract, and also contains a contract glossary, while the workbook tests the reader's estimating skills and helps to identify the strengths and weaknesses of their businesses.

The second handbook and workbook start where the first set finishes – a potentially profitable contract has been won. The first part of these books, "planning for profit", helps the reader to prepare a realistic plan to carry out the work, while the second part, "making it happen", deals with the principles and practice of site supervision.

The third handbook and workbook cover business management. A contracting firm is not just a collection of individual contracts; it is also a business enterprise. These books focus on financial control and office administration, which are frequently neglected by contractors who are generally more interested in the technical aspects of building work.

The way the IYCB system works is that the *handbook* provides ideas and information and the *workbook* gives the reader a chance to look at his or her business in a disciplined way, and decide on action plans to make it more competitive and successful. Together, the IYCB series should enable you, as the owner or manager of a small construction enterprise, to improve *your* construction business. As joint authors, with between us about a hundred years' experience of working with small contractors around the world, we understand the risky and demanding environment in which you work and hope that the IYCB series will help you and your firm to survive and prosper.

This book was prepared and edited under the auspices of the ILO's Construction Management Programme, which was initiated within the Entrepreneurship and Management Development Branch of the Enterprise and Cooperative Development Department, and is now based in the Policies and Programmes for Development Branch of the Employment and Development Department.

Claes-Axel Andersson

Derek Miles

Richard Neale

John Ward

THE AUTHORS

Claes-Axel Andersson manages the Improve Your Construction Business project within the ILO Construction Management Programme, which is based in its Policies and Programmes for Development Branch. He is a professionally qualified civil engineer with extensive experience in project management and building design.

Derek Miles is Director of Overseas Activities in the Department of Civil Engineering at the Loughborough University of Technology, United Kingdom. He is a Fellow of the Institution of Civil Engineers and the Institute of Management and has more than 20 years' experience in the development of national construction industries. He directed the ILO Construction Management Programme during the period 1986-94.

Richard Neale is Senior Lecturer in the Department of Civil Engineering at the Loughborough University of Technology, United Kingdom. He is a professionally qualified civil engineer and builder, and is a consultant to the ILO and other international organizations in construction training and development.

John Ward is an independent consultant specializing in training for construction enterprises, and was previously chief technical adviser to the first Improve Your Construction Business project. He started his career as site agent and engineer with major construction companies, then ran his own small contracting business before specializing in the training of owners and managers of small construction enterprises.

ACKNOWLEDGEMENTS

The *Improve Your Business (IYB)* approach to small enterprise development was conceived by the Swedish Employers' Confederation, and has since been developed internationally by the ILO with financial assistance from the Swedish International Development Authority (SIDA) and other donors.

The Government of the Netherlands agreed to finance the first "Improve Your Construction Business" (IYCB) project, based at the Management Development and Productivity Institute (MDPI) in Accra.

Ghana proved a good choice. As a result of recent changes there is a more favourable climate for private sector initiatives, and Ghanaians have a well-deserved reputation for entrepreneurial drive. The Civil Engineering and Building Contractors Association of Ghana (CEBCAG) appreciated the opportunity that the project offered for its members to improve their management skills, and worked closely with the MDPI team and the ILO Chief Technical Adviser to ensure that the training programme met the most urgent needs of its members

This initial IYCB project provided an opportunity to develop and test a series of *Improve Your Construction Business* handbooks and workbooks and we wish to specifically acknowledge the dedication and enthusiasm of the MDPI/CEBCAG training teams or "cohorts".[1] The project package contained a certain amount of material that was specific to operating conditions in Ghana, but this published edition has been carefully edited to meet general needs of owners and managers of small-scale construction enterprises for basic advice on ways to improve business performance.

[1] Yahaya Abu, Michael Adjei, Margaret Agyemang, Kofitse Ahadzi, Henry Amoh-Mensa, Ernest Asare, John Asiedu, Franklin Badu, Fidelis Baku, Siegward Bakudie, Joseph Dick, Hamidu Haruna, Mathias Kudafa, D. Nsowah, Eric Ofori, Yaw Owusu-Kumih, S. Sakyi, Harry Seglah.

CONTENTS

HOW TO USE THIS HANDBOOK

This handbook is written for *you* – the owner or manager of a small construction business. Together the three basic IYCB handbooks provide advice on most aspects of running such a business, and the three complementary workbooks give you the chance to test your management skills, assess the performance of your business in a disciplined way and develop your own action plans.

Improve Your Construction Business is material for you to work with. It is available in a series of modules which take you step-by-step through the different stages of running a small contracting business. They are best read together. We suggest you first read the chapter in the handbook, and then work through the examples in the corresponding chapter of the workbook.

This handbook

This handbook contains a worked example of a simple building project, showing how to plan a typical project using bar charts, labour schedules, materials schedules, etc. It is both a basic textbook and a reference book, showing how you can plan your projects using a step-by-step approach. The chapters are set out in the same order as the chapters in the workbook, so that you can easily go from workbook to handbook or from handbook to workbook.

The workbook

The workbook enables you to test your planning skills by means of exercises in management practice. It will also make you think hard about how you can make your company more profitable through improved productivity on your sites by asking you a number of questions.

In each chapter of the workbook there is a list of simple questions to which you answer "yes" or "no". Your answers will tell you about the strengths and weaknesses of your business.

If you find that you need to improve your management skills in certain areas after going through the workbook, you can turn back to the appropriate section in the handbook and make sure you understand all the items and techniques introduced there.

Where to start

The following route map will help you to find your way around the handbook. We recommend that you start by reading quickly through the whole book. Then you can go back over it more slowly, concentrating on the chapters which deal with those parts of management which you think are weakest in your business.

Planning for profit				Making it happen			
Choice of technology	Allowables	Bar charts Labour schedules Plant and transport schedules Materials schedules	Checking on progress	Supervision Site layout	Productivity Improving work methods Incentive schemes	Health and safety	Quality control
Discusses the choice of technology and the use of equipment, describes the use of allowables to provide time or cost targets, the preparation of bar charts and schedules for labour, plant, transport and materials and how to check on progress.				Describes how to supervise a construction site and how to choose an efficient site layout, explains the importance of **productivity** and how to improve work methods, describes various types of incentive schemes, outlines health and safety procedures and emphasizes the importance of **quality control**.			
Chapter 1	Chapter 2	Chapter 3 Chapter 4 Chapter 5 Chapter 6	Chapter 7	Chapter 8 Chapter 9	Chapter 10 Chapter 11 Chapter 12	Chapter 13	Chapter 14

As soon as you feel comfortable with the ideas in a particular chapter, you can try out your skills in the workbook. Together this handbook and workbook, and the others in the IYCB series, should become your "business friends".

> Note: Since this book is intended for use in many different countries, we have used the term "NU" in the examples to represent an imaginary "National Unit of currency" and NS to stand for imaginary "National Standards".

2

PART A
PLANNING FOR PROFIT

CHOICE OF TECHNOLOGY   I

This handbook is about the management of construction projects – in other words it is about how you should get work done
on your construction sites. One of the special features of the
construction industry is that it offers a wide choice of possible
technologies – ways of getting things done. Clients are increasingly interested in how you carry out the work as well as the end
result. Since businesses only exist as long as they please their
clients, this should also interest you. As a contractor or manufacturer, you need to make the right choice of technology if you
are to carry out your projects effectively and profitably. So we
suggest that you read this short chapter before going on to the
technical aspects of planning and site management.

Adaptability

Sometimes the choice of technology is decided by the client, by
specifying the method which the contractor should use as well
as the end result. This limits the choice available to the contractor, but it can also provide a market opportunity for those contractors who can adapt quickly to new ways of doing things.
Adaptability is a key construction skill, so this first chapter has
been written to help you ensure that your business has the flex-
ibility to learn new skills and change techniques to suit changes
in the market for the services that you provide.

APPROPRIATE
CONSTRUCTION TECHNOLOGY

Governments are the main client for construction works in most
countries, and they are also in a position to influence designs by
private clients through setting contract laws and building regulations. Increasingly they are exercising this power to ensure that

the technology used by the construction industry is appropriate, which means that the limited natural resources available are used in a sensible and economic way and that there is a proper balance between the use of human skills and mechanical plant and equipment.

In developing countries where there is a severe shortage of foreign exchange and a need to create new employment opportunities for school-leavers and people displaced from traditional jobs in the rural areas, it is not surprising that governments are keen to encourage local resource-based technologies. This means designing projects so that maximum use is made of local materials and so that they can be constructed by local people without having to rely on expensive imported equipment.

As far as the government is concerned, these policies result in the output of useful roads, buildings and public works projects, but they also help cope with national social and economic problems. Even if you take the view that national social and economic problems are not your main concern, you need to be aware of these trends or you will lose your business to contractors who have adapted more quickly to the changing needs of the market.

LOCAL RESOURCE-BASED TECHNOLOGIES

What are the implications of local resource-based technologies for your business? The trend towards using labour rather than plant and equipment means that you will have to become better at managing people. People are often more difficult to manage than plant. You have to listen to them and treat them fairly. You have to think about how they will react to changes in the way you run your business. But one good definition of management is "getting things done with and through people". If you cannot manage people and you call yourself a contractor, you are probably in the wrong business.

The increasing emphasis on the use of local resources could bring business opportunities as well as problems. There will be good business opportunities in the small-scale manufacture of building materials. Microconcrete roof tiles are a good example of a new technology that can be operated economically on a small scale without a heavy initial investment in plant and equipment. The trend towards appropriate construction technology will favour small businesses generally, since the big contractors and large-scale manufacturers will lose their competitive advantage.

WHAT EQUIPMENT TO BUY

Traditionally contractors were often graded according to the amount of equipment they possessed. In the future they will be graded mostly according to their record and the skills of their workforce. You may be in a better position to compete for road contracts if you own tractors and trailers rather than expensive imported earthmoving equipment. In a changing market, the best equipment to own is equipment which is versatile, and which will be useful whatever type of project you obtain. You should concentrate on owning good basic equipment such as concrete mixers, and investing enough in maintenance and spare parts so that it is always ready for service.

SCAFFOLDING AND FORMWORK

Conservation of natural resources means that there will be restrictions on the waste of scarce timber supplies. This will mean a trend away from the use of wooden poles for scaffolding and timber for formwork. Even though steel scaffolding and formwork is expensive to buy (and is not produced on a small scale), it can be a good investment because it can be re-used many times.

PLANT HIRE

Another market opportunity that will emerge as a result of changing technologies is in the hire of specialist plant. This chapter has suggested that it may no longer pay for contractors to own large and expensive items of equipment. But specialist equipment will still be needed from time to time, and contractors will be willing to pay good rates for hiring it because it will still be cheaper than having their own equipment lying idle in their plant yard when it is not required.

Their problem could be your opportunity. If you are good at buying, operating and maintaining plant, it may pay to add a specialized plant hire service to your contracting business. In many countries, most contractors have their own plant hire subsidiary companies and all their plant is expected to "earn its keep", either being hired out to the contractor's own sites at a preferential internal hire rate or hired to outsiders at the full rate.

ALLOWABLES   2

Definition

Allowables are the costs or time periods that a contractor can "allow" for the job, covering labour, plant and transport, and still make the planned profit.

These allowables will help you to plan the job more efficiently as you can use them for cost control and planning.

A standard set of labour output records is available in many countries and is used to set average "norms" of labour output. These norms, which give the average unit time (in workdays or workhours) for each operation, are called "labour constants".

However, as a professional contractor you should be able to calculate "allowables" from the direct project costs yourself.

Allowables are calculated "backwards" from estimating data, a bill of quantities, received.

CALCULATING ALLOWABLES

The procedure for calculating and using allowables will be carried out step-by-step following an example. The example is from the construction of three buildings shown on pages 101 to 103. There is also a direct project costs chart for these buildings on the following pages.

Step 1   Calculating cost allowables

The cost allowables are calculated by dividing the direct project cost for an item of work by the total quantity of work to be done on that item.

Answers to business practice - 6

1. THE DIRECT PROJECT COSTS CHART

			DIRECT PROJECT COSTS CHART					
	List of quantities taken off drawings				Direct project costs (NU)			
Item No.	Description	Unit	Quantity	Labour	Plant	Materials	Transport	Total
1.	Clear site	–	–	20	–	–	10	30
2.	Excavate top soil	m²	300	125				125
3.	Excavate foundations	m³	75	375				375
4.	Steel to foundations 12 mm 8 mm	m	900 216	234		622	7	863
5.	Formwork to foundations	m²	54	156		118	10	284
6.	Concrete to foundations	m³	12.0	84	24	309	60	477
7.	Steel to columns 12 mm 8 mm	m	693 228	312		495	7	814
8.	Formwork to columns	m²	147	312		245	15	572
9.	Pour concrete to columns	m³	11.1	336	96	284	54	770
10.	Concrete block walls up to floor	m²	96	279		1 196	90	1 565
11.	Return fill and ram excavated mtl. around found.	m³	51	117				117
12.	Hardcore fill	m³	51	252	72	372	31	727
13.	Mesh to floor	m²	153	156		324	4	484
14.	Concrete to floor	m³	20.4	504	144	524	92	1 264
15.	Concrete block walls above floor	m²	102	279		1 280	99	1 658
16.	Soffit forms to ring beam over openings	m²	9	78		15	2	95
17.	Soffit forms over infill panels	m²	7	78		12	2	92
18.	Sideforms to ring beam	m²	54	234		90	5	329
19.	Supply and fix steel 12 mm to ring beam 8 mm	m	432 162	234		315	7	556
20.	Pour concrete to ring beam	m³	7.8	336	96	201	36	669
21.	Fabricate roof trusses	No.	33	312		733	20	1 065
22.	Fix roof trusses	No.	33	156		51	3	210
						Total forward (points 1-22)		13 141

DIRECT PROJECT COSTS CHART								
List of quantities taken off drawings				Direct project costs (NU)				
Item No.	Description	Unit	Quantity	Labour	Plant	Materials	Transport	Total
23.	Roof tile battens	m	636	39		140	5	184
24.	Tile roof (incl. ridge)	m²	243	207		2 881	50	3 138
25.	Timber to gable end	m²	18	78		132	5	215
26.	Form eaves Horizontal and gable ends	m	108	78		146	2	226
27.	Supply and fix ceiling boards	m²	126	54		555	15	624
28.	Prefabricated window panels	No.	12	08		900		1 008
29.	Prefabricated door panels	No	3	54		300		354
30.	Terrazzo floor	m²	132	108	30	300	4	442
31.	Plaster walls and columns and ringbeam	m²	144	156		75	18	249
32.	Paint walls and ceilings	m²	270	126		405	15	546
33.	External paths and parking	m²	146	115	40	340	79	574
34.	Spread topsoil to landscape site	m²	300	60				60
35.	Perimeter fence	m	115	92		316	10	418
36.	Dispose of surplus material off site	m³	65	80			49	129
				Final total of direct project costs				21 308

Example Item 6: Pour concrete to foundations
Direct project costs to mix and place 4.0 m³ of concrete to foundations:

Labour = 28 NU/4.0 m³ = 7.00 NU/m³
Plant = 8 NU/4.0 m³ = 2.00 NU/m³
Transport = 20 NU/4.0 m³ = 5.00 NU/m³

If you get figures that are difficult to use when calculating (e.g. 3.54 NU/m²), the allowables can be rounded off:

Labour cost allowable = 7 NU/m³
Plant cost allowable = 2 NU/m³
Transport cost allowable = 5 NU/m³

Step 2   Calculating time allowables

We use the **cost** allowable to calculate a **time** allowable.

The cost for a concrete gang is calculated to 28 NU/0.5 days, that equals:

Actual labour cost of concrete gang = 56 NU/day (28/0.5 = 56)

Labour cost allowable = 7 NU/m^3

Time allowed for concrete gang to mix and place 1 m^3 unit of concrete is labour cost allowable divided by labour cost per concrete gang.

That gives you: 7 NU divided by 56 NU/day = 0.125 days.

Time allowable = 0.125 days/m^3 = 1 hr/m^3.

Step 3   Allowables chart

In order to go through all the allowables methodically, it is wise to use an allowables chart. Cost and time allowables are now entered on the following allowables chart:

Item	Description	Unit	Cost allowables			Time allowables		
			Labour	Plant	Transport	Labour	Plant	Transport
3.	Pour concrete to foundations	m^3	7	2	5	Concrete gang 0.125 days	–	–

USING ALLOWABLES

The allowables calculated are used to work out the most cost-effective time period in which the contract can be completed. They will also help us in Chapters 3 and 4 when we prepare bar charts and labour schedules to make our planning more efficient.

Step 4
Calculating most cost-effective time period

Example:
If there is a total of 100 m^3 of concrete foundations to be poured, and the labour time allowable is 0.125 day/m^3, then the most cost-effective time period in which foundation con-crete can be completed is:

$$100 \times 0.125 = \underline{12.5 \text{ days}}$$

Remember to make sure that your equipment can keep pace with the labour.

When working continuously a 5/3.5 mixer normally pro-duces about 10 m^3 of concrete per day (1.2 m^3 per hour). To produce 1.0 m^3 of concrete would therefore take 1.0/10 = 0.10 day (less than time allowable for labour = 0.125 day). The capac-ity of the mixer is superior to the projected labour output.

Labour is the determining factor since it takes the concrete gang about 12 days to complete the task while the capacity of the mixer is higher (10 days).

In this example, we have only calculated the allowable related to labour but we have also checked that the capacity of our plant is sufficient. The transport element of this item has not been calculated since it is not directly related to the task (all transportation can be done well in advance and therefore does not need to influence the output). In some cases, for instance, if you are using ready-made concrete, the transport capacity can determine the possible output.

Experienced contractors often recommend calculating the size of equipment you need and then ordering one size larger. This is to make sure you have an extra margin of output if you run into problems resulting in reduced production. This extra output can help you back on the track again and make it poss-ible to meet your production targets even if you experience some setbacks.

BAR CHARTS   3

Purpose

Bar charts are pictures which tell us when and how the work is going to be done. The bar chart for a job is the immediate result of your planning and tells you when the different operations are going to start, when they will be finished and how they fit in with each other. You can use it to estimate when to order construction material and equipment, when there is need for extra manpower, etc. Paradoxically, a very important use of these charts is to help you when the work does not proceed as planned and you have to make changes; the chart then gives you a possibility to assess the likely consequences of a change; for instance, what other activities will be affected, and how, if an activity is delayed two weeks.

Preparing a bar chart is simple, but preparing a realistic and cost-saving bar chart can be a bit more difficult. The only way to master this technique is to learn the theory properly and then practise it several times.

One good reason for putting some extra effort into your planning is that consultants often judge contractors by how realistic the accompanying bar chart is when tendering.

On page 16 there is an example of a bar chart. As a contractor you might receive a chart like this from the client. Here you can see how your construction work fits in with all the other phases of the project. You have five months to finish your work (August to December):

BAR CHART SUPPLIED BY CLIENT

Item	Jan	Feb	Mar	Apr	May	Jun	Jul	Aug	Sep	Oct	Nov	Dec
1. Briefing	▬											
2. Sketch plans		▬										
3. Working drawings			▬	▬	▬	▬						
3a. Initial estimates			▬	▬	▬							
3b. Statutory permissions			▬	▬	▬	▬						
4. Tender documents						▬						
5. Tendering							▬					
6. Construction								▬	▬	▬	▬	

HOW TO MAKE A BAR CHART

Now we are going to learn how to prepare a bar chart that describes our part of the project, the construction phase, by splitting it into activities.

There are six major steps that go into making a bar chart:

1. Plan
2. List jobs
3. Calculate quantities
4. Calculate time
5. Draw the bar chart
6. Check.

Plan

Start by going through the project, step by step, from the very beginning to the last activity, in your mind. Make sure that you really think about all the different stages. Sometimes you have to split complex activities down into smaller units.

List jobs

Write down all the operations to be carried out on the project. This will help you to work out how much labour, equipment, material, etc., is going to be needed (this work could already have been done when taking off from the drawings in preparing a list of quantities).

Examples: Excavate foundations

Pour concrete to foundations

Tile roof.

Calculate quantities

The number of workers and type of equipment determine how long the operations will take. The duration of each operation has to be known in order to programme the project. Calculating the quantities and calculating the time, step 4, can be done in one operation if the work is not very complicated.

The key point is that the duration depends on the size (and skills) of the workforce.

Calculate time

The duration of each job in days or months will be used to programme the project. This can be done using the previously calculated time-allowables (Chapter 2), combined with the contractor's own valuable experience, knowledge and common sense. A job that requires 10 workweeks can theoretically be done by:

<table>
<tr><td></td><td>1 worker in</td><td>10 weeks</td></tr>
<tr><td>or</td><td>2 workers in</td><td>5 weeks</td></tr>
<tr><td>or</td><td>5 workers in</td><td>2 weeks</td></tr>
<tr><td>or</td><td>10 workers in</td><td>1 week.</td></tr>
</table>

What is the best solution? The two extremes (1 worker in 10 weeks, 10 workers in 1 week) are often ridiculous alternatives while the best solution probably lies somewhere in the middle (between 2 workers in 5 weeks and 5 workers in 2 weeks). The best solution depends on the situation, so you have to use your experience when deciding.

Draw the bar chart

The horizontal bars, each representing a job, are drawn on the bar chart when the best starting and finishing times have been decided. Think carefully about how different jobs are linked to each other. Remember to use a pencil, not a pen, since the bar chart will be changed several times during the planning process.

Questions to ask yourself:

When is the earliest time the operation can start?

What other operations must be finished before this one can start?

What overlaps (operation starting after another operation has started but before it is finished) can be allowed with other operations?

What other operations can be done at the same time?

Check

When the chart is finished check that no mistakes have been made and whether there is scope for any improvements. These bar charts should always be made together with labour schedules and materials schedules since they influence each other very much.

Remember: A bar chart is not a decoration to put on your office wall. A continuing process of updating must take place regularly during the project to adjust your chart to changing circumstances and to let you know whether your work is on schedule.

WHAT TO LOOK FOR

When you have started to make your bar chart, remember that the most important source of information is your own experience. You know how many workers you need in a team, what different jobs can be done simultaneously, how much longer it takes to finish a task if the conditions are bad, etc. Bar charts must never simply be an exercise with figures. Always look at your chart and ask yourself if it is the best possible solution.

Ask yourself:

Will there be enough labour in that period?

Will the equipment be available in that period?

Will the materials be available then?

Are there likely to be any problems with working space, transportation of materials, etc., in that period?

Have I made the best use of my resources?

If your answers to these questions are satisfactory, you are on your way to preparing an efficient programme for the project.

Preparing a bar chart – an example:

The construction bar chart shows whether the manager has chosen the best possible starting and finishing times for the operations. The following example of a bar chart is based on the list of quantities[1] taken off the working drawings at the end of the book, on pages 101 to 103. It shows how the construction work can be completed within five months, or twenty weeks. In order not to make this bar chart too detailed and complicated, the shortest time-limit used is half a week.

❑ Two items, consisting of similar activities that are undertaken together, are shown as one bar only, especially if the activities are very short. Examples: Items 16-18 "Formwork to ring beam" cover soffit forms and sideforms; Items 28-29 "Prefabricated panels" cover both window and door panels.

❑ Some activities can be overlapped, e.g. once the foundations for one building have been completed and the excavation gang moves on to the next building then Item 4 "Steel to foundations" and Item 5 "Formwork to foundations" can start on building number one.

❑ Note that Item 6 "Concrete to foundations" can start when the reinforcement and the formwork (Items 5 and 6) are completed on the first building. However, since it is a very short activity, the concrete gang would then have to wait for the formwork and reinforcement on the next house to be ready. It is up to the manager to decide whether it is advantageous to undertake the activity in this way or to wait until all three houses can be done one after the other, without a break. The same is true for Item 20, "Concrete to ringbeam".

❑ The manager has allowed extra time for the concrete to cure before putting on any loads. For example, the ring beam needs to be cured before starting with Item 22 "Fix roof trusses".

---

[1] How to prepare a complete list of quantities is shown in *Pricing and bidding (IYCB 1) Handbook*, Chapter 5.

BAR CHART - CONSTRUCTION PHASE

Item	Week number																			
	01	02	03	04	05	06	07	08	09	10	11	12	13	14	15	16	17	18	19	20
1-2. Clearing and excavating top soil	‖																			
3. Excavate foundations		‖																		
4. Steel to foundations			‖																	
5. Formwork to foundations				‖																
6. Concrete to foundations					‖															
7. Steel to columns						‖														
8. Formwork to columns							‖													
9. Concrete to columns							‖													
10. Block wall, up to floor								‖												
11. Return fill and ram								‖												
12. Hardcore fill									‖											
13. Mesh to floor									‖											
14. Concrete to floor										‖										
15. Block wall, above floor											‖									
16-18. Formwork to ringbeam												‖								

Activity	01	02	03	04	05	06	07	08	09	10	11	12	13	14	15	16	17	18	19	20
19. Steel to ringbeam											▌	▌								
20. Concrete to ringbeam													▌							
21. Fabricate roof trusses														▌						
22. Fix roof trusses															▌					
23. Roof tile battens																▌				
24. Tile roof																▌	▌			
25. Timber to gable ends																	▌			
26. Form eaves																▌				
27. Ceiling boards																▌				
28-29. Prefabricated panels																	▌			
30. Terrazzo floor																	▌	▌		
31. Bagwash walls, columns																		▌		
32. Paint																			▌	▌
33. Paths and parking																		▌		
34. Spread topsoil																		▌		
35. Fence																			▌	
36. Surplus material off-site																				▌
Week no.	01	02	03	04	05	06	07	08	09	10	11	12	13	14	15	16	17	18	19	20

❑ On some items the manager has allowed extra time for the concrete to cure before other activities are allowed to start although they do not involve any loads. Item 10 "Block wall, up to floor" awaits the columns; Items 28-29 "Prefabricated panels" need to await the ring beam.

❑ Items 33-36 do not depend on items related to the work undertaken inside the building, although they are related to each other. These activities can therefore start while other activities are undertaken inside the building.

❑ Although no labour schedule has been prepared yet, you can already start to note down alternative solutions, such as what can be prefabricated. Parts of both the steelfixers' and the carpenters' tasks can often be done in advance.

❑ In this case nothing is done by subcontractors. When sub-contractors are used, you do not need to plan their activities in detail, but you must make sure their work fits in with the other activities enabling them to perform their tasks. For example, make sure that painters can have access to the rooms to be painted.

LABOUR SCHEDULES 4

Purpose

A labour schedule shows what workforce is needed and when it should be on site. It is mainly based on the bar charts we learned how to prepare in Chapter 3 and on your experience as a builder. The main factor to consider when deciding on the number of workers on site is that it should be kept as even as possible throughout the project and with minimal gaps between activities for each worker. This is achieved by a continuing exchange between labour schedules and bar charts where you have to adjust both of them in order to meet this goal.

HOW TO MAKE LABOUR SCHEDULES

The labour schedules are drawn up using the bar charts already prepared. You write down the number of workers in each category that will be needed for each activity. We will learn by going through an example, based on the bar charts from Chapter 3. In the example the skilled and general labourers have been separated. The item and description on the labour schedules are the same as on the bar chart. The number of workers required semi-weekly could have been worked out when calculating the direct project cost and should have been considered when preparing the bar chart. This number may have to be changed in order to get an even distribution. As with the bar charts, the labour schedules should always be made using a pencil, not a pen, to facilitate changes. Once again, the best source of knowledge for preparing a labour schedule is that already inside your head – it is called practical experience.

SEMI-WEEKLY LABOUR SCHEDULE – SKILLED WORKERS

Item	Week number																			
	01	02	03	04	05	06	07	08	09	10	11	12	13	14	15	16	17	18	19	20
1-2. Clearing and excavation of topsoil																				
3. Excavate foundations																				
4. Steel to foundations		2	2 2	2															steelfixers	
5. Formwork to foundations			2 2	2 2															carpenters	
6. Concrete to foundations				2 2	2													mixer operator vibrator operator		
7. Steel to columns					2 2	2 2													steelfixers	
8. Formwork to columns					2	2 2	2												carpenters	
9. Concrete to columns						2 2	2 2											mixer operator vibrator operator		
10. Block wall, up to floor							2 2	2 2											masons	
11. Return fill and ram							1	1 1	1										any skilled	
12. Hardcore fill								1 1	1 1									roller operator		
13. Mesh to floor									2 2	2									steelfixers	
14. Concrete to floor										2 2	2 2	2						mixer operator vibrator operator		
15. Block wall, above floor										2	2 2	2							masons	
16-18. Formwork to ringbeam											4	4 4							carpenters	
19. Steel to ringbeam											2	2 2	2						steelfixers	
20. Concrete to ringbeam													2 2	2 2				mixer operator vibrator operator		
21. Fabricate roof trusses												2 2	2 2	2 2					carpenters	

Multi-resource / activity schedule chart (weeks 01–20). Split cells shown as "a|b"; dashed lines shown as em-dashes in the original.

Activity / Resource	01	02	03	04	05	06	07	08	09	10	11	12	13	14	15	16	17	18	19	20
22. Fix roof trusses															2|2	2			carpenters	2|2
23. Roof tile battens																I	I		carpenters	
24. Tile roof										tilers						I	I			
25. Timber to gable ends										carpenters					I	I				
26. Form eaves										carpenters						I				
27. Ceiling boards										carpenters					I	I				
28-29. Prefabricated panels										carpenters							2|2			
30. Terrazzo floor											float operator					I		I		
31. Bagwash walls, columns										masons								I	I I	
32. Paint										painters								I		
33. Paths and parking											roller operator							I		
34. Spread topsoil																				
35. Fence										carpenters								I	I I	
36. Surplus material off-site																				
Steelfixers		2	2|2	2|2	2|2	2|2			2|2	2	2	2	2							
Concrete gang			2|2	2|2	2	2|2	2|2		2	2|2	2	6|6	2|2							
Carpenters			2|2	2|2	2|2	2|2	2			2		4	2|2	2|2	2|2	5 4	3 2			
Roller operator								I I	I								I			
Masons							2|2	2|2		2	2|2	2								
Float operator																				
Tiler																		I		
Painters									I								I	I I	I	2|2
Any skilled								I I	I										I I	
Week no.	01	02	03	04	05	06	07	08	09	10	11	12	13	14	15	16	17	18	19	20

Examples of questions related to the labour schedule that you should ask yourself are:

Is there an uneven spread of workforce between the weeks?

Can I change the order in which the items are done to get a more even spread?

Can the idle workforce be employed to prepare another item such as prefabrication of components, or can they help another gang?

Look at the labour schedule on pages 24 to 25.

It is now possible to check the spread of the workforce over the period covered by the contract. The skilled labour schedule shows an uneven spread, with some idle time:

❑ The steelfixers are idle for half a week at the end of week 4, all of weeks 7-8 and for one week in weeks 10 and 11.

❑ The mixer operator and the vibrator operator are idle for half a week in week 5 and for 1.5 weeks in weeks 8-9 and 11-12.

❑ The carpenters are idle for half a week in week 5, during weeks 7-11 and during week 18. In week 11 and week 16 we need, with the current bar chart, 4-6 carpenters who should not be hired for such a short period of time.

❑ The masons are idle for 1.5 weeks during weeks 9-10 and for 5.5 weeks during weeks 12-17.

❑ Item 11, marked "any skilled", is supervision of labourers that can be done by any of the skilled workgroups.

The manager should now study possible ways of keeping the skilled workforce busy and to even out high demand, such as:

❑ The work order can in some cases be changed, notably for items that are *not* related to the completion of other items. Examples:

Items 33-36 are not really related to the completion of the inside work (although they are in some cases related to each other) and can therefore be done when suitable labour is available;

Items 25-26 are only related to previous items (the roof trusses have to be fixed) but not to the items following, i.e. they can also be done when they fit in with available labour.

❑ During their idle weeks the steelfixers could:
1. move to another site
2. prefabricate the steel for the columns and the ring beams and prepare the mesh for the floor
3. work with the concrete gang.

- The carpenters and the mixer and vibrator operators could be used as gang bosses for the labourers.

- The carpenters could prefabricate forms during week 6 and roof trusses during weeks 8, 9 and 10 to reduce the need for carpenters in weeks 11 and 12 and start items 25 and 26 "Gable ends" and "Form eaves" earlier to reduce the need for carpenters in weeks 16 and 17.

- If available, you can of course always allocate additional labour to enable you to finish work earlier.

- You can also allocate fewer workers than originally en-visaged if the completion of the item does not interfere with other items or if you can start the activity earlier and still finish it on time.

- If the number of general labourers hired makes it possible to better utilize the skilled labour, i.e. reduce idle time, the extra cost for the labourers is often lower than what is gained by reducing the number of workdays for skilled personnel.

Before making any changes we will therefore look at the labour schedule for general labourers (pages 28 to 29) as well.

The general labourers' schedule shows a rather uneven spread, with some idle time and some periods when there is a shortage of labourers.

It is, of course, not practical to have one worker on site one week, five the next, 12 the third and fourth weeks, eight on the fifth week, and so on. It is better to even out by deciding upon the number of workers required and replan the work in such a way that you can keep that number fairly steady throughout the project.

- When replanning the work to get a more even spread, you need, of course, to compare your labour schedule for general labourers with the one for skilled labour and your bar chart since they all depend on each other. One way to get a good picture of the labour requirements, both for general labourers and for skilled labour, is to draw a "Labour needed" schedule as shown on page 30. That enables you to see easily the time periods when you will have problems related to the number of labourers on site.

- The manager may decide to aim at starting with eight labourers (weeks 1-2), build up to 12 for the majority of the contract, then transfer four to another site towards the end of the contract (weeks 14-20), bringing the workforce down to eight.

SEMI-WEEKLY LABOUR SCHEDULE - GENERAL LABOURERS

Item	Week number																			
	01	02	03	04	05	06	07	08	09	10	11	12	13	14	15	16	17	18	19	20
1-2. Clearing and excavation of topsoil	5 ¦ 5																			
3. Excavate foundations	5	5 ¦ 5	5 ¦ 5	5																
4. Steel to foundations		¦ 2	2 ¦ 2	2																
5. Formwork to foundations			2 ¦ 2	2 ¦ 2																
6. Concrete to foundations				8 ¦ 8	8															
7. Steel to columns					2 ¦ 2	2 ¦ 2														
8. Formwork to columns					2	2 ¦ 2	2													
9. Concrete to columns						8 ¦ 8	8 ¦ 8													
10. Block wall, up to floor							3 ¦ 3	3 ¦ 3												
11. Return fill and ram							1	1 ¦ 1	1											
12. Hardcore fill								4 ¦ 4	4 ¦ 4											
13. Mesh to floor									2 ¦ 2	2										
14. Concrete to floor									8	8 ¦ 8	8									
15. Block wall, above floor										3	3 ¦ 3	3								
16-18. Formwork to ringbeam											4	4 ¦ 4								

Activity	01	02	03	04	05	06	07	08	09	10	11	12	13	14	15	16	17	18	19	20
19. Steel to ringbeam											2	2	2							
20. Concrete to ringbeam												8	8							
21. Fabricate roof trusses												2	2	2						
22. Fix roof trusses															2	2				
23. Roof tile battens																1				
24. Tile roof																3	3			
25. Timber to gable ends																1	1			
26. Form eaves																1	1			
27. Ceiling boards																1				
28–29. Prefabricated panels																	3	3		
30. Terrazzo floor																	2	2		
31. Bagwash walls, columns																		1	1	1
32. Paint																				1
33. Paths and parking																		3		
34. Spread topsoil																		4		
35. Fence																			3	
36. Surplus material off-site																				4
Total	5 10	5 7	9 7	17 10	10 4	12 12	13 12	8 8	7 14	10 11	11 11	11 8	12 10	2 2	2 2	5 7	7 11	10 7	4 4	5 5
Week no.	01	02	03	04	05	06	07	08	09	10	11	12	13	14	15	16	17	18	19	20

For comments, see pages 27 and 31.

LABOUR NEEDED

It is often very useful to show the need for labour graphically.

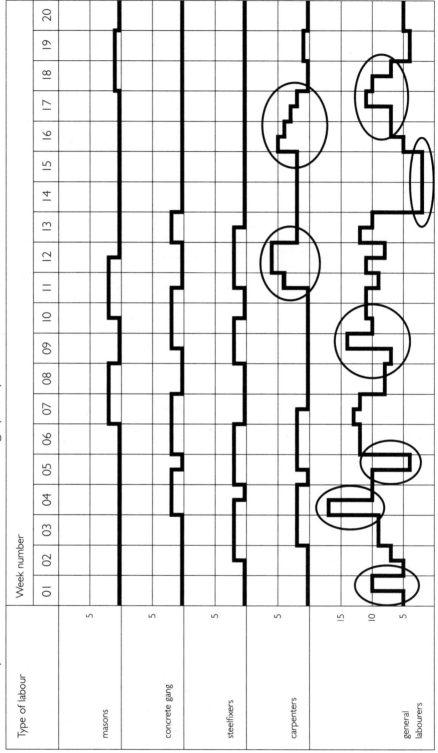

When looking at the "Labour needed" schedule on the previous page you can easily identify where there is an uneven spread of labour.

❏ The *masons* are needed during weeks 7 and 8 and 10-12, resulting in a gap in weeks 9 and 10. One mason is needed in weeks 18 and 19. Could the masonry work scheduled for weeks 10-12 (item 15) start earlier?

❏ The *mixer operator* and the *vibrator operator* have three periods of idleness; half a week in week 5, 1.5 weeks in weeks 8 and 9 and 1.5 weeks in weeks 11 and 12. Could these gaps be avoided by starting item 9 and item 20 earlier than planned? Since it does not seem possible to bring all concrete work together, it is often better to create a substantial gap between the activities, enabling you to transfer the concrete gang to another site. Another alternative is to use them as gang bosses for the labourers, as mentioned earlier.

❏ The *steelfixers* show a very similar pattern to the concrete gang. They also have three shorter periods of idleness: half a week in week 4, two weeks during weeks 7 and 8 and one week during weeks 10 and 11. Some of these gaps could be bridged by prefabricating steelworks, thereby also shortening the time needed later, or alternatively you could try to bring activities together, as for the concrete gang.

❏ With the *carpenters* we have two sorts of problem: some periods of idleness and other periods when we would need up to six carpenters at the same time. Half a week is idle in week 5. A period of four weeks' idleness (weeks 7-11) is followed by a need for up to six carpenters during weeks 11 and 12. A second period of very high demand for carpenters (up to five) occurs in weeks 16 and 17. There is also a gap in week 18, followed by a demand for one carpenter. Could we prefabricate formwork during the idle period in week 5 or maybe start item 8 earlier? Item 21, "Fabricate roof trusses", can be done during weeks 7-11. The same is true of the carpentry work needed on item 35, "Fence". Items 25 and 26 are not really holding up any other activities, i.e. you can schedule them to fit in with an even distribution of carpenters. Also remember that although you might have planned to undertake an activity quickly with a large number of workers you can allocate fewer workers if you can start earlier, and still be able to complete in time.

❑ When looking at the *general labourers'* schedule, there are six periods when it is difficult to keep to the plan to start with eight, then go up to 12 and later bring it down to eight again, as mentioned earlier: week 1 where we start with five and then immediately go up to 10; week 4 with a momentary need for 17 labourers; week 5 when we go down to four labourers; week 9 when there is a need for 14; weeks 14 and 15 when we need only two; and weeks 16, 17 and 18 when we need 11 labourers.

Some problems, such as week 1, can be solved by starting with eight labourers, which should enable us to complete items 1-2 earlier and thereby avoid the need for 10 labourers later on. A "gap" of half a week, as in week 5, when we need significantly fewer labourers is not really a problem since it is often useful to have a buffer. If some of the activities for weeks 16-18 could be done during weeks 14 and 15, we could even out the need for labourers. Remembering the need to transfer activities that we identified on the carpenters' chart, it seems that we could make this change.

Finding a good solution to all these problems of uneven spread, for both skilled labour and general labour, could sometimes be rather time-consuming. It is, however, definitely worthwhile since it is much cheaper to make mistakes at this stage than on the site. This is not an exercise in mathematics but a practical planning task that must be based on your experience. Remember:

PLANNING = HIGHER PRODUCTIVITY = HIGHER PROFIT

After carefully looking at alternative solutions the manager has settled for the plans and the bar chart shown on the following pages. Which are the main changes from what was originally planned?

❑ The two main tasks of the *masons* have been brought together so that they start with item 15, "Block wall, above floor", immediately after finishing item 10, "Block wall, up to floor". Undertaking item 15 at the same time as item 14, "Concrete to floor", will need careful planning by the manager. Although these items are not really dependent on each other, the workers could easily get in each other's way. The masons will have to be ready before concreting starts in a house otherwise they will probably damage the newly concreted floor.

❑ The *mixer operator* and the *vibrator operator* now work more or less continously from week 4 to week 12 (inclusive). Item 9, "Concrete to columns", now starts immediately after item 6, "Concrete to foundations", is finished.

These two operators have been assigned to the activity that could be done by any of the skilled workers: item 11.

Item 20, "Concrete to ring beam", now starts a week earlier than originally envisaged, thereby reducing the gap to item 14, "Concrete to floor", to a mere half-week.

❑ The *steelfixers* can start prefabricating steel for the columns in the second half of week 4 and thereby finish item 7 half a week earlier than originally planned and close the gap we had in week 4. The following activities will, of course, also be able to start half a week earlier.

The starting date for item 19, "Steel to ring beam" is brought closer to the anticipated finishing date of item 13, "Mesh to floor", to create another longer period of continuous work for the steelfixers. The idle period in weeks 6-8 has deliberately been kept long enough to enable a temporary transfer to another site.

❑ The demand for *carpenters* has been evened out so that we now have an almost constant need for two carpenters from week 3 to 18. Item 8, "Formworks to columns", follows the move of item 7 and starts half a week earlier.

Items 16-18 were originally planned to be undertaken by four carpenters during 1.5 weeks. Now it will take longer (2.5 weeks) to complete this task since only two carpenters have been assigned, but because of an early start the activity will be completed half a week earlier than originally envisaged.

The carpenters start fabricating the roof trusses, item 21, during the otherwise idle weeks 7-9 and finish that job with an additional week just before item 22, "Fix roof trusses", starts.

Items 25 and 26, "Timber to gable ends" and "Form eaves", will be undertaken later than originally planned to even out the demand for carpenters, since these activities are not related to others.

"Prefabricated panels", items 28-29, will be done during 1.5 weeks instead of one week as originally planned.

The carpentry work on item 35, "Fence", is done in advance, during week 12, to even out the need for carpenters.

Item	01	02	03	04	05	06	07	08	09	10	11	12	13	14	15	16	17	18	19	20	Skilled worker
1-2. Clearing and excavation of topsoil																					
3. Excavate foundations																					
4. Steel to foundations		2	2 2	2														steelfixers			
5. Formwork to foundations			2 2	2 2														carpenters			
6. Concrete to foundations				2 2	2													mixer operator / vibrator operator			
7. Steel to columns				2	2 2	2												steelfixers			
8. Formwork to columns					2 2	2 2												carpenters			
9. Concrete to columns					2	2 2	2											mixer operator / vibrator operator			
10. Block wall, up to floor						2	2 2	2										masons			
11. Return fill and ram							2	1 1									any skilled work done by concrete gang				
12. Hardcore fill							1	1 1	1									roller operator			
13. Mesh to floor								2	2 2									steelfixers			
14. Concrete to floor									2	2 2	2							mixer operator / vibrator operator			
15. Block wall, above floor								2	2 2	2								masons			
16-18. Formwork to ringbeam										2 2	2 2	2						carpenters			
19. Steel to ringbeam										2	2 2	2						steelfixers			
20. Concrete to ringbeam												2 2						mixer operator / vibrator operator			
21. Fabricate roof trusses							2	2 2	2			2	2 2					carpenters			

Activity	01	02	03	04	05	06	07	08	09	10	11	12	13	14	15	16	17	18	19	20
22. Fix roof trusses														2¦2	2			carpenters		2¦2
23. Roof tile battens															1 1	1 1		carpenters		
24. Tile roof									tilers							1 1	1 1			
25. Timber to gable ends									carpenters								1			
26. Form eaves									carpenters									2¦2		
27. Ceiling boards									carpenters							1	1 1			
28-29. Prefabricated panels									carpenters							2	1 1			
30. Terrazzo floor										float operator									1	
31. Bagwash walls, columns									masons					1				1 1	1 1	
32. Paint									painters											
33. Paths and parking										roller operator										
34. Spread topsoil																				
35. Fence									carpenters			2								
36. Surplus material off-site																				
Steelfixers		2	2	2¦2	2¦2	2		¦2	¦2	2	2¦2	2								
Concrete gang				2¦2	2¦2	2¦2	2¦	2	2	2¦2	2¦1	2¦2								
Carpenters			2¦2	2¦2	2¦2	2¦2	¦2	2¦2	2	2¦2	2¦2	2¦2	2¦2	2¦2	3¦2	1¦2	2¦2	2		
Roller operator								1	1	2			1 1							
Masons							2	1	1				1							
Float operator												2								
Tiler								1¦1							1 1	1 1	1¦1	1	1	
Painters															1 1	1 1	1¦1	1 1	1 1	2¦2
Any skilled							¦2	1¦1	1¦1											
Week no.	01	02	03	04	05	06	07	08	09	10	11	12	13	14	15	16	17	18	19	20

For comments, see page 33.

SEMI-WEEKLY LABOUR SCHEDULE – GENERAL LABOURERS (REVISED)

Item	Week number																			
	01	02	03	04	05	06	07	08	09	10	11	12	13	14	15	16	17	18	19	20
1-2. Clearing and excavation of topsoil	8 | 5																			
3. Excavate foundations	3 |	8 | 6	8 | 8																	
4. Steel to foundations		2 |	2 | 2	2 |																
5. Formwork to foundations			2 | 2	2 | 2																
6. Concrete to foundations				8 | 8	8 |															
7. Steel to columns				2	2 | 2	2														
8. Formwork to columns					2 | 2	2 | 2														
9. Concrete to columns					| 8	8 | 8	8 |													
10. Block wall, up to floor						3	3 | 3	3												
11. Return fill and ram							2	1 | 1												
12. Hardcore fill							4 |	4 | 4	4 |											
13. Mesh to floor								2	2 | 2											
14. Concrete to floor									8 |	8 | 8	8 |									
15. Block wall, above floor								3 |	3 | 3	3										
16-18. Formwork to ringbeam										2 | 2	2 | 2	2								

Activity	01	02	03	04	05	06	07	08	09	10	11	12	13	14	15	16	17	18	19	20
19. Steel to ringbeam										2	2	2								
20. Concrete to ringbeam											8	8								
21. Fabricate roof trusses							2	2	2				2	2						
22. Fix roof trusses								2	2					2	2					
23. Roof tile battens															1	1				
24. Tile roof															3	3	3			
25. Timber to gable ends																	1	1		
26. Form eaves																		2		
27. Ceiling boards																1	1			
28-29. Prefabricated panels																2	1	1		
30. Terrazzo floor																	2	2		
31. Bagwash walls, columns																	1	1	1	
32. Paint																			1	1
33. Paths and parking													2	2						
34. Spread topsoil																		2	2	
35. Fence														2				2	2	
36. Surplus material off-site																			4	4
Total	8	8	12	12	12	12	11	10	11	13	12	12	4	4	3	4	4	5	5	5
Week no.	01	02	03	04	05	06	07	08	09	10	11	12	13	14	15	16	17	18	19	20

For comments, see page 33.

LABOUR NEEDED (REVISED)

Type of labour	Week number																			
	01	02	03	04	05	06	07	08	09	10	11	12	13	14	15	16	17	18	19	20
masons	5																			
concrete gang	5																			
steelfixers	5																			
carpenters	5																			
general labourers	15 10 5																			

The rest of the activity, erecting of the fence, can be done by general labourers.

Concerning the result in weeks 15 and 16 (I) as shown on the "Labour needed" schedule: three carpenters are needed for half a week in the begining of week 15, while only one carpenter is needed at the end of week 16.

Item 23, "Roof tile battens", does not start until item 22, "Fix roof trusses" is ready (as a little too much labour has been allocated). Items 28-29, "Prefabricated panels", can start slightly earlier with one carpenter doing preparations.

❑ The *roller operator* is still scheduled to work at the site during two distinct periods, weeks 7-9 and 13-14. There is a possibility to programme item 33 to follow the completion of item 12, but we also need to have two labourers available and the allocated workforce of 12 labourers is already set aside for other activities during that period. However, there might be a chance to reprogramme during the course of the project; if an activity is finished earlier than originally planned, the manager should keep this reprogramming option in mind.

❑ The labour-needed chart shows a much more even spread of *general labourers* than the original version. We can even reduce the number of labourers needed during the last eight weeks of the project to five instead of the originally anticipated eight. We can now start with eight labourers during the first two weeks, then build up to 12 during weeks 3-12 and bring them down, as previously mentioned, to five during the remainder of the project.

Items 1-2, "Clearing of site" and "Excavation of topsoil" will be done by eight labourers instead of five to enable us to complete earlier.

Item 3, "Excavate foundations", will also be done by a larger number than originally planned, making it possible to complete that activity in a shorter period of time.

"Formwork to ringbeam", items 16-18, is allocated a lower number of workers and will thereby take longer to complete. See also above concerning the carpenters.

Items 34-36 are programmed to give us a constant need for labourers during the last weeks of the contract since these activities are not related to the last activities undertaken inside the building. The labour-needed schedule still gives the impression that more than 12 labourers are needed

Item	Week number																			
	01	02	03	04	05	06	07	08	09	10	11	12	13	14	15	16	17	18	19	20
1-2. Clearing and excavation of topsoil	▌																			
3. Excavate foundations		▌	▌																	
4. Steel to foundations		▌	▌	▌																
5. Formwork to foundations			▌	▌																
6. Concrete to foundations				▌	▌															
7. Steel to columns				▌	▌															
8. Formwork to columns					▌	▌														
9. Concrete to columns					▌	▌	▌													
10. Block wall, up to floor						▌	▌													
11. Return fill and ram							▌	▌												
12. Hardcore fill								▌	▌											
13. Mesh to floor									▌											
14. Concrete to floor									▌	▌	▌									
15. Block wall, above floor								▌		▌										
16-18. Formwork to ringbeam										▌		▌								

Gantt chart — construction schedule (weeks 01–20)

Task	01	02	03	04	05	06	07	08	09	10	11	12	13	14	15	16	17	18	19	20
19. Steel to ringbeam										█										
20. Concrete to ringbeam													█							
21. Fabricate roof trusses							█	█												
22. Fix roof trusses															█					
23. Roof tile battens															█					
24. Tile roof																█				
25. Timber to gable ends																	█			
26. Form eaves																		█		
27. Ceiling boards																█				
28–29. Prefabricated panels																█				
30. Terrazzo floor																		█		
31. Bagwash walls, columns																			█	
32. Paint																				█
33. Paths and parking														█						
34. Spread topsoil																		█	█	
35. Fence									carpenters preparing			█		█					█	
36. Surplus material off-site																				█
Week no.	01	02	03	04	05	06	07	08	09	10	11	12	13	14	15	16	17	18	19	20

in week 6 (II) and weeks 9 and 10 (III). However, item 10, "Block wall, up to floor", will not start until item 8, "Formwork to columns", is ready.

To ensure an even distribution during weeks No. 9 and 10, we will not start item 14, "Concrete to floor", until item 13, "Mesh to floor", is ready and items 16-18, "Formwork to ringbeam", will await the completion of item 15, "Block wall, above floor".

PLANT AND TRANSPORT
SCHEDULES 5

Purpose

A plant and transport schedule shows you what equipment items are needed and when they should be on site. It can also help you to decide whether to buy or hire plant or transport. In this way plant and transport utilization can become more efficient. Plant and transport schedules are generally planned to fit in with the bar chart that you have already prepared. However, if very costly plant has to be used, you may have to alter the chart again so that this costly plant is on site for as short a time as possible.

HOW TO MAKE PLANT
AND TRANSPORT SCHEDULES

When preparing plant and transport schedules we use the bar charts already made. We will learn by going through an example based on the revised bar chart from Chapter 4. The plant and transport weekly requirements should have been worked out at the direct project cost and bar chart preparation stages. As with bar charts and labour schedules, remember that the best source of knowledge for preparing a plant and transport schedule is the contractor's own experience.

Some of the main factors to consider when deciding when items of plant will be required are:

❑ Is there a technical reason why the job needs to be done by an item of plant? Example: compaction of soil can often only be done properly by plant.

PLANT AND TRANSPORT SCHEDULE

Item	Week number																			
	01	02	03	04	05	06	07	08	09	10	11	12	13	14	15	16	17	18	19	20
1-2. Clearing and excavating topsoil	‖																Truck			
3. Excavate foundations		‖																		
4. Steel to foundations			‖													Truck (all reinforcement steel)				
5. Formwork to foundations			‖													Truck (for all the woodwork)				
6. Concrete to foundations				‖														Mixer, Vibrator, Bowser		
7. Steel to columns																				
8. Formwork to columns																				
9. Concrete to columns					‖	‖												Mixer, Vibrator, Bowser		
10. Block wall, up to floor																				
11. Return fill and ram																				
12. Hardcore fill							‖		‖								Pedestrian roller			
13. Mesh to floor																				
14. Concrete to floor										‖	‖						Mixer, Vibrator, Bowser			
15. Block wall, above floor																				
16-18. Formwork to ringbeam																				
19. Steel to ringbeam																				
20. Concrete to ringbeam												‖						Mixer, Vibrator Bowser		

Gantt chart — construction schedule (weeks 01–20)

Activity	Equipment / notes	Bar (week range)
21. Fabricate roof trusses	Circular saw	08–09
22. Fix roof trusses		
23. Roof tile battens		
24. Tile roof		
25. Timber to gable ends		
26. Form eaves		
27. Ceiling boards		
28–29. Prefabricated panels		
30. Terrazzo floor	Power float	17–18
31. Bagwash walls, columns		
32. Paint		
33. Paths and parking	Pedestrian roller	13–14
34. Spread topsoil		
35. Fence	Circular saw; Concrete plant	12; 14; 17
36. Surplus material off-site	Truck	
Concrete plant		04–07; 09
Circular saw		09; 12; 14
Roller		09; 13–14
Truck		01–02
Power float		18

Week no.: 01 02 03 04 05 06 07 08 09 10 11 12 13 14 15 16 17 18 19 20

For comments, see next page.

- Is it more economical to do the job by an item of plant, or is a labour-intensive method more efficient?
- Is plant available where and when it is needed?
- Will deliveries be made to the site by the supplier or is it more economical for the contractor to supply transport and collect materials from the factory, workshop or quarry?
- Is there a cheap, efficient public transport service available to bring labour to the site or is it more economical to use your truck to provide transport?

Look at the plant and transport schedule on pages 44 to 45. It is now possible to check at a glance the distribution of plant and transport over the period covered by the contract:

- The concrete plant is utilized for seven weeks out of the 11 weeks it will spend on the site. That equals 64 per cent usage, which is quite good. However, could we avoid the gaps? During the last week, week 14, it is only used for the fence. Maybe that could be done earlier for instance in week 11 or 13.
- Do we need to have gaps of several weeks in the utilization of the circular saw and the pedestrian roller?

The manager should now study possible ways of utilizing the plant and transport during idle periods:

- It seems difficult to avoid a gap between item 9, "Concrete to columns", and item 14, "Concrete to floor" and maintain an acceptable level of utilization. Do we have another site where the concrete equipment could be utilized?
- The second time that the circular saw is required (when the carpenters start fabricating the fence, item 35) it is only needed for half a week. We should try to undertake this job during the first period (item 21) during weeks 6-9 so that the saw can be sent away afterwards.
- Maybe item 33, "Paths and parking", could be done after item 12 is finished, so as to avoid the gap in the utilization of the roller (and the roller operator)? However, since the paths run alongside the buildings and a number of activities concerning the outside and the roof of the building would go on during this work and after its completion there is a risk of having to do some complementary work afterwards or even redo some of the work.

❑ .Maybe we should have a hoist on site when tiling the roof? We must remember that moving plant on and off the site is expensive and it would only be needed for a very short period. We have a tiler and three labourers assigned to this task but during the week when we have programmed the tiling there should be some additional labourers available. It is probably more economical to stick to a "labour-intensive" way, but you should always check alternatives.

Remember always to think through all alternatives available for completing a task and to calculate the differences carefully, including all related costs. It is often easy to forget the cost of transport, erection and dismantling of plant.

Monthly hire of plant almost always results in a lower daily rate than weekly hire, so sometimes the potential loss because of plant standing idle for a short period of time can be compensated by a lower daily rate.

After making changes to the plant and transport schedule, the manager should check how they affect the labour schedule and the bar chart. If necessary, these should be altered until the best solution is found. In this way you can achieve even more efficient labour, plant and transport utilization, and significant cost savings.

Remember that planning is a job that you can never hope to get right the first time. It is better to spend time trying out ideas at the planning stage when the only equipment you are using is paper and a pencil, than to wait until the construction stage and lose money through wasting the costly time of employees and equipment.

MATERIALS SCHEDULES 6

Purpose

A materials schedule shows the contractor what materials are needed and when they should be on site. In addition to acting as a guide for ordering materials, the schedule also serves as a checklist of materials needed for the project. It is usually minor items that get forgotten and cause temporary delays and disorganization. As a checklist, the materials schedule helps you to avoid such problems.

HOW TO MAKE MATERIALS SCHEDULES

When preparing materials schedules we use the bar charts already made. A materials schedule normally contains the following information:

❑ What is to be ordered

❑ How much to order

❑ When it will be required

❑ Which part of the building or which item on the list of quantities the materials are for.

It may also be convenient to include information such as:

❑ The name of the supplier

❑ The date on which to order the materials.

The information needed to make the materials schedule comes from two main sources:

1. The quantities are taken from the materials calculation for tender.

2. The dates for which the materials are required are taken from the bar chart, and the ordering dates depend on how

Information obtained from calculations of quantities				Date that materials are needed on site (from bar chart)	Time needed between order and delivery (information at planning stage)	Latest date that order must be placed	Details of supplier					Remarks
Item	Description	Unit	Quantity				Order no.	Name	Address	Phone	Contact	
4, 7, 13, 19	Reinforced steel	kg	2 500	Week 2 Aug. 8th	2 weeks	July 18	161	J. Doe	High Street	2242	John	Allows 30 days credit
5, 8, 16 - 18, 21 - 23, 25 - 26, 35	Wood 1" x 4" 1" x 1" 1.5" x 6" 1.5" x 4" 2" x 4" 1.5" x 8" stakes 4" x 4"	m m m m m m m	2 650 700 120 450 425 40 100	Week 3 Aug. 15th	2 weeks	July 25	171	Pine Ltd.	Easy Street	1845	Mr. Fibre	Cash with order
6,9	Concrete to foundations and columns Aggregate Sand Cement	kg kg kg	29 000 17 400 7 800	Week 4 Aug. 22nd Aug. 22nd	1 week 2 weeks	Aug. 8 Aug. 1	177 176	Crash quarries Opee cement	Ridge quarry Cement works	2962 2229	Mr. Crash Mr. Opee	Cash with order Cash on delivery

Information obtained from calculations of quantities				Date that materials are needed on site (from bar chart)	Time needed between order and delivery (information at planning stage)	Latest date that order must be placed	Details of supplier					Remarks
Item	Description	Unit	Quantity				Order no.	Name	Address	Phone	Contact	
10, 15	Concrete blocks	No.	2 900	Week 6 Sep. 5th	3 weeks	Aug. 8	182	AYS Ltd	West Industrial Area	2127	Mrs. Ambee	Cash with order
14, 20	Concrete to floor											
	Aggregate Sand	kg kg	35 100 21 300	Week 9 Sep. 26th	1 week	Sep. 12th	179	Crash quarries	Ridge quarry	2962	Mr. Crash	Cash with order
	Cement	kg	9 600	Sep. 26th	2 weeks	Sep. 5th	178	Opee cement	Cement works	2229	Mr. Opee	Cash on delivery
24	Tile roof	m²	243	Week 15 Nov. 7th	2 weeks	Oct. 17th	201	A.N. Other Tile Works	Industrial area	No phone	Mr. Other	90 days credit; 20% cash discount

In order to give us time to check and if necessary correct the material on delivery, the latest day for ordering has been set one week earlier.

long the suppplier takes to deliver the materials. If the sup- plier takes one week to deliver, then order the materials at least a week before they are needed.

In addition to delivery time, allow for the materials to arrive on site a few days before they are needed. This gives time for stacking and preparation, and makes it possible to correct mis- takes in the delivery if it does not exactly match the order.

In the example shown on pages 50 to 51, the "item" and "description" columns are the same as on the revised bar chart that we have used in previous examples. Unless there is a big difference between the contract and working drawings, the materials requirements will be available from the calculations that were worked out at the "taking-off" stage. To show the principle we have noted the first six orders in the materials schedule.

CHECKING ON PROGRESS 7

Purpose

By recording how the construction of an item progresses, you will be able to see whether it will be completed on time. If it looks as though the project will be delayed it is better to discover this at the earliest possible moment so that action can be taken to get the project back on schedule. It is always easier to correct a problem or mistake at an early stage.

HOW TO CHECK ON PROGRESS

On the following pages there are examples of one method of recording progress. In these examples, a line is drawn down the bar chart at the end of the week. Everything on the left of the line is work that has been completed and the bars are filled in to show actual progress. The programme situation can be seen at a glance.

Example I shows the programme situation at the end of week 5:

❑ Item 7, "Steel to columns", has just started, although it should be almost complete.

❑ Item 8, "Formwork to columns", is three-quarters complete, which means that the carpenters are pushing hard against the steelfixers.

❑ "Concrete to columns", item 9, has not been started since the steelfixers are not ready with house no. I.

❑ In order to keep the carpenters working, the site foreman has told them to start fabricating the roof trusses, so approximately half a week of that job has been completed, as marked on bar "Fabricate roof trusses", item 21.

EXAMPLE 1 . BAR CHART - CONSTRUCTION PHASE

End of week no. 5

Item	Week number																			
	01	02	03	04	05	06	07	08	09	10	11	12	13	14	15	16	17	18	19	20
1-2. Clearing and excavating top soil																				
3. Excavate foundations																				
4. Steel to foundations																				
5. Formwork to foundations																				
6. Concrete to foundations																				
7. Steel to columns																				
8. Formwork to columns																				
9. Concrete to columns																				
10. Block wall, up to floor																				
11. Return fill and ram																				
12. Hardcore fill																				
13. Mesh to floor																				
14. Concrete to floor																				
15. Block wall, above floor																				
16-18. Formwork to ringbeam																				

Task	01	02	03	04	05	06	07	08	09	10	11	12	13	14	15	16	17	18	19	20
19. Steel to ringbeam										▮										
20. Concrete to ringbeam												▮								
21. Fabricate roof trusses					▭	▭	▪		▮											
22. Fix roof trusses													▮							
23. Roof tile battens															▮					
24. Tile roof															▮					
25. Timber to gable ends																▮				
26. Form eaves																	▮			
27. Ceiling boards																▮				
28-29. Prefabricated panels																▮				
30. Terrazzo floor																	▮			
31. Bagwash walls, columns																			▮	
32. Paint																				▮
33. Paths and parking														▮						
34. Spread topsoil																		▮		
35. Fence												▮		▮					▮	
36. Surplus material off-site																				▮

End of week no. 10

EXAMPLE 2

Item	Week number																			
	01	02	03	04	05	06	07	08	09	10	11	12	13	14	15	16	17	18	19	20
1-2. Clearing and excavating top soil																				
3. Excavate foundations																				
4. Steel to foundations																				
5. Formwork to foundations																				
6. Concrete to foundations																				
7. Steel to columns																				
8. Formwork to columns																				
9. Concrete to columns																				
10. Block wall, up to floor																				
11. Return fill and ram																				
12. Hardcore fill																				
13. Mesh to floor																				
14. Concrete to floor																				
15. Block wall, above floor																				
16-18. Formwork to ringbeam																				

Task	01	02	03	04	05	06	07	08	09	10	11	12	13	14	15	16	17	18	19	20
19. Steel to ringbeam																				
20. Concrete to ringbeam																				
21. Fabricate roof trusses																				
22. Fix roof trusses																				
23. Roof tile battens																				
24. Tile roof																				
25. Timber to gable ends																				
26. Form eaves																				
27. Ceiling boards																				
28-29. Prefabricated panels																				
30. Terrazzo floor																				
31. Bagwash walls, columns																				
32. Paint																				
33. Paths and parking																				
34. Spread topsoil																				
35. Fence																				
36. Surplus material off-site																				
Week no.	01	02	03	04	05	06	07	08	09	10	11	12	13	14	15	16	17	18	19	20

EXAMPLE 3

End of week no. 15

Item	Week number																			
	01	02	03	04	05	06	07	08	09	10	11	12	13	14	15	16	17	18	19	20
1-2. Clearing and excavating top soil																				
3. Excavate foundations																				
4. Steel to foundations																				
5. Formwork to foundations																				
6. Concrete to foundations																				
7. Steel to columns																				
8. Formwork to columns																				
9. Concrete to columns																				
10. Block wall, up to floor																				
11. Return fill and ram																				
12. Hardcore fill																				
13. Mesh to floor																				
14. Concrete to floor																				
15. Block wall, above floor																				
16-18. Formwork to ringbeam																				

	01	02	03	04	05	06	07	08	09	10	11	12	13	14	15	16	17	18	19	20
19. Steel to ringbeam																				
20. Concrete to ringbeam																				
21. Fabricate roof trusses																				
22. Fix roof trusses																				
23. Roof tile battens																				
24. Tile roof																				
25. Timber to gable ends																				
26. Form eaves																				
27. Ceiling boards																				
28-29. Prefabricated panels																				
30. Terrazzo floor																				
31. Bagwash walls, columns																				
32. Paint																				
33. Paths and parking																				
34. Spread topsoil																				
35. Fence																				
36. Surplus material off-site																				
Week no.	01	02	03	04	05	06	07	08	09	10	11	12	13	14	15	16	17	18	19	20

Example 2 shows the programme situation at the end of week 10:

❏ "Mesh to floor", item 3, is only half complete and the steel fixers have not yet started with item 19, "Fix steel to ring beam", while the carpenters have more or less completed items 16-18, "Formwork to ring beam"; they have also almost completed item 21, "Fabricate roof trusses", and they have done their part of item 35, "Fence".

❏ Item 14, "Concrete to floor", is of course also delayed since the steelfixers are not ready with the mesh.

❏ It should be obvious by now that the bar chart shows an unbalanced system of working – the steelfixers need either help or motivation to catch up with the carpenters; otherwise there is a risk that the entire project may be delayed.

Example 3 shows the programme situation at the end of week 15:

❏ Due to reasons that became obvious in examples 1 and 2, the site foreman had to ensure that the steelfixers could catch up; it was possible to transfer an additional steelfixer from another site for two weeks and two additional labourers were also assigned to the steelfixers.

❏ Because fewer labourers have been assigned to the carpenters they have lost some of their advance but they are still ahead of schedule.

❏ Now we have a balanced system of working with strong indications that we will be able to finish the project on time or even ahead of schedule.

The future can never be predicted accurately, and that is certainly true of construction projects. When you plan a project, you know that not everything will happen in the way you predicted. People not turning up for work, machine breakdowns, bad weather and delayed payments are all examples of things which may happen suddenly and delay the project. Such incidents are realities of life and have to be coped with as they arise. Therefore, planning has to be flexible to adapt to these realities, which means that planning has to go on continuously throughout the project, it is *not* a one-time operation.

PART B
MAKING IT HAPPEN

SUPERVISION 8

Planning for supervision

Proper supervision is necessary for a workforce to operate efficiently. It is needed at various levels, but it is particularly important at site agent and site foreman level.

Proper site supervision is needed for the following reasons:

❑ To motivate the workers to work efficiently

❑ To make sure the quality of the work is up to standard

❑ To ensure that safety and security regulations are followed, both as regards the workforce and the general public

❑ To maintain a high activity level

❑ To have people on the job who can report on problems

❑ To give the workforce clear instructions

❑ To have people on site who can suggest more efficient ways of doing things

❑ To authorize payment to the workers

❑ To measure productivity.

There are five important factors to note when arranging for proper site supervision:

1. Allocate the right number of workers per supervisor

2. Organize your workforce at all levels and make sure that everybody knows the structure of the organization

3. Establish systems of reporting

4. Recruit effective supervisors

5. Assign each supervisor a clearly defined area of responsibility and give supervisors authority equal to their responsibility.

1.   Allocate the right number
of workers per supervisor

If there are too many workers per foreman, it becomes difficult to keep an eye on them all. If there are too few, it means that your supervision costs will be high. Start with the number of workers per supervisor which, from experience, seems right. Then observe, as the work progresses, how well the individual supervisors cope with the workers allocated to them.

2.   Organize at all levels

All supervisors should know exactly who and what they are responsible for and who they are responsible to.

The foremen supervise the gang leaders. The site manager supervises the foremen. The contractor (or the manufacturer) supervises the site manager.

If the site staff at all levels know that they have to answer for the progress of those under their supervision, they have an additional incentive to supervise well.

3.   Establish systems of reporting

Every supervisor must report daily to his superior on the operations and daily activities which he is responsible for. Reporting should be done at every level where there is a supervisor. The foreman, for example, should prepare daily reports, describing what has taken place during the day. This will help him to remember all the information which he is to give to the site manager and the contractor (or the manufacturer). The daily report also contains information which will be useful to check up on later.

Example of a daily record: This should be prepared by the same person <u>every</u> day

Job: Concrete to foundations	Date: 23.09.96		
Description of job	Labour used		
Mixing and placing concrete to strip footings. Building No. 1	Trade	Number	Hours
	Mixer operator	1	6
	Mixer gang	2	6
Delays to job	Wheelbarrow gang	4	6
	Vibrator operator	1	5
Delays due to having to clean out dirt from footings which had fallen from piles of excavated material along trench	Placing gang	2	6
	Clearing foundations	2	5

4. Recruit effective supervisors

The contractor's prime responsibility is to recruit good, qualified foremen and site managers, who can be trusted to get the work done quickly and to an accepted standard of quality. You depend on them. A good site manager can ensure that a project is executed without cost or time overruns. A poor site manager can cause great losses to the contractor throught high site costs and long delays. A very bad site manager can even lead the company into bankruptcy. Loyalty should be properly rewarded, so the entrepreneur should seriously consider paying extra for good site staff, with bonuses for increased productivity.

A *site manager* is responsible for running the operations on site efficiently. He must be able to prepare daily and weekly programmes, keep records, coordinate the workforce efficiently and be a good trouble shooter. It is important for a site manager to have an interest in executing the project *as quickly* and *as well* as possible.

The site manager should generally be regarded as a member of the permanent staff, so he will be motivated to get jobs finished quickly without worrying about "working himself out of a job". Even so, motivation can be reinforced and monetary rewards should be considered if he succeeds in executing the project promptly.

A *foreman* is closer to the workers. Whereas the site manager needs to have good overall administrative abilities, the foreman needs to possess personal qualities which lead the workers to respect him and to work as efficiently as possible. These are some qualities which the foreman should possess: honesty; integrity; leadership; authority; capacity to organize; ability to motivate others and to maintain discipline.

It is important to train site managers and foremen to carry out their supervisory duties effectively.

5. Authority must match responsibility

It is important that a supervisor at any level is given the authority to match the responsibility. In the world of sport, it is the captain's responsibility to make sure that the football team never goes into action without each player knowing who he is supposed to mark and what the goalkeeper and the striker are meant to be doing, and yet, in the world of work, many contractors neglect to take similar precautions in their business.

Problem solving

Problems can mount up very quickly on a site, in a workshop, or in a fabrication yard. They should be tackled on a regular basis, as a problem solved at an early stage will usually take less of the contractor's time than one which is left to get worse.

How do you learn how to take problems in their logical sequence and find good solutions? Here is a suggestion that we have found to be very useful.

Step 1   List the problems

In order to solve the problems, the contractor must first identify them. The best way to do this is to sit down and list them as they come to mind, as shown below with lines of varying length to indicate the relative importance of each problem:

```
A    ————————————————————
B    ——————————————————————
C    ————————————
D    —————————————————————
E    ——————————
F    ————————
G    ———————————————
H    —————————————————
```

Step 2   Grade the problems

The contractor should study each of the problems to estimate which is the most serious, and grade them in order of seriousness. Is there a problem which, if solved, will lead to the solving of others?

When laid out in order of seriousness or difficulty, the problem list will look like this:

```
B    ——————————————————————
D    —————————————————————
H    —————————————————
A    ————————————————————
G    ———————————————
C    ————————————
E    ——————————
```

Step 3   List all possible solutions

Each problem should then be considered carefully, starting with the most serious. Every possible solution should be listed and followed through until the most logical solution can be found.

Here is an example of what a list like this can look like:

Problem: The concrete gang's productivity is low.

Possible solutions	For	Against
1. Sack the old gang and train a new gang.		* Takes too long to train new gang. * Trouble with unions.
2. Buy a bigger mixer.	* Should improve concrete output.	* Old mixer is giving ample output. * New mixer is very expensive.
3. Study the method of worksite layout to see if productivity can be improved or the size of the gang cut down.	* Can be delegated to section foreman. * No extra cost. * Gang can be put on piecework or bonus. * Morale should improve. * It will impress the client.	* If no improvement is made, the section foreman will have wasted his time.

Step 4   Solve the problems

The entrepreneur should then take steps to organize practical methods of solving the problems. Can responsibility for solving some of them be delegated? Can some problems be left to a later stage?

In the previous example the obvious solution is to delegate responsibility for solving the problem and bring in changes immediately.

Problems when making the site more efficient

Raising productivity can be seen as a threat by the workers, because some productivity measures lead to workers being laid off if they are found to be redundant. This dissatisfaction can lead to problems on the site, in the workshop, or in the fabrication yard. If your workers feel that their jobs or income are threatened, they may protest or sabotage your attempts to increase productivity. This can apply to supervisors as well as workers.

There are two main reasons why supervisors might decide not to support your productivity measures:

- If it is revealed that operations for which they are responsible can be greatly improved, it will show that they have not been doing a satisfactory job. Everybody is proud of their work, including the supervisor. Therefore, new methods or systems which are introduced can threaten their pride.

- If disputes arise as a result of activities to improve productivity, they are the persons who will be most affected.

Nothing breeds suspicion like attempts to hide the purpose of what is being done or the likely effects. Where measures are intended to improve the conditions for the workers and the health of the company, there is nothing to hide. Point out that, as the firm becomes more competitive and they become more effective, more contracts should be won, offering continuity of employment and prospects of promotion.

To end this section on problem solving, here is a summary of the advices:

- ❏ Problems should not be ignored since they will not go away by themselves
- ❏ List the problems – don't just keep thinking and worrying about them
- ❏ Tackle the problems in a positive way
- ❏ Tackle the worst problem first: it may be that solving it will help towards solving the rest
- ❏ Always inform the supervisors and workers of changes to come
- ❏ Do not be afraid to ask others for suggestions on how to solve problems.

SITE LAYOUT 9

Good site layout is important to improved productivity. You need to arrange your construction site in such a way as to make it function as efficiently as possible.

With bad layout, time and materials are wasted through double handling. Transport and handling of materials always cost money. Every time you move a stack of bricks around your site, the real cost to you increases. One of the reasons for slow progress and high cost of construction projects is the lack of site planning, including poor site layout.

The layout of the site will depend upon two main factors: the methods and sequence of operations to be employed in carrying out the work, and the space available. The methods and sequence will have been considered at the time that the estimate and tender were prepared. When the contract is awarded, you should take another careful look at these ideas before preparing a detailed site layout plan.

Ask yourself the following questions:

❑ How can I cut handling and stacking time?

❑ How can I reduce the distances that materials and workers have to travel?

❑ Are the piles or stacks of material located close to where they are to be used?

❑ Are materials properly stacked for ease of storage and handling?

Generally, the site will have to accommodate a variety of temporary buildings, materials and supplies, plant and equipment at different times. Schedules will be needed giving their respective times of arrival and departure from the site. Particular care should be taken to avoid items blocking access and interfering with the activities at the various stages of a project. It is essential to list all the items and storage areas that will be needed on site, and to locate their position on a site plan.

Good layout is most important when the product or materials being used are heavy or big, as in woodworking, reinforcing steel assembly, or concrete precasting. In wood machining, the machines cut very fast. If the machines are in the right order there is little delay between the stages of manufacture, and the movement of the timber is reduced to a minimum.

Without a precise site layout plan, neither the site manager nor other site staff will have a clear indication of where stores and offices, items of plant, work areas, and stacks of materials should be located. Then the following may happen:

1. Materials stores may be wrongly located

2. Fixed plant and equipment may be wrongly located

3. Not enough space may be allowed for stacking and preparation

4. Temporary buildings may be wrongly located.

Materials stores wrongly located

Materials arriving on site are unloaded into what someone guesses to be the correct location. This practice may subsequently involve double or triple handling of materials to move them to another place because, for example:

❑ They have been stacked over a drainage run, or in the way of the scaffolding, or too near the edge of a future excavation

❑ They are too far from where they are needed

❑ They are too far from the hoist, or not within the radius of the crane, or a long way from the ramp

❑ They obstruct work traffic across the site

❑ They are too near the route of work traffic and may get damaged or soiled

❑ Their delivery was wrongly timed and they are not needed until much later in the project; if the materials concerned are expensive and/or fragile, this is even more serious, since they are likely to get damaged or stolen if they are left lying about for a long time.

Fixed plant and equipment wrongly located

Mixer:
- ❑ inaccessible for delivery of materials
- ❑ not enough room for storage of aggregates
- ❑ wrongly located for fast delivery of mixed concrete.

Hoist:
- ❑ insufficient capacity or height in relation to the loads to be handled, or the nature of the building
- ❑ badly located in relation to the floor layout of the building.

Not enough space allowed for stacking and preparation

- ❑ Materials may consequently have to be stacked too high or may intrude into roadways or other areas, presenting hazards and causing breakages
- ❑ Working areas become too cramped for efficiency, or additional areas have to be allocated, with consequent waste of time travelling between them.

Temporary buildings wrongly located

Temporary buildings may be wrongly located in relation to their effective use and convenience, such as:

Site office:
- ❑ too near noisy activities such as mixer or carpenter's shop
- ❑ too near site or roads in dusty conditions
- ❑ too remote to give a sufficient overview of the site.

Stores:
- ❑ inadequate access for unloading/loading
- ❑ causes dangerous practices when loading/unloading, thereby putting the safety of the workforce in danger
- ❑ cement store too far away.

Latrines:
- ❑ located upwind of the office
- ❑ located in badly drained areas.

To avoid such problems, it is necessary to prepare a detailed site plan on which every item of accommodation and equipment as well as ancillary work areas and material storage areas are located. On a cramped or complicated site, a series of plans will be needed covering the layout at each stage of the work.

These plans must be prepared in advance and they must be prepared by someone with good planning skills and experience. It is often easier to save money through planning than by raising site productivity, but these savings can only be obtained if you take enough <u>time</u> to think about all the possible alternatives before deciding on how to lay out the site.

The table below lists some important considerations which you should take into account when deciding on site layout.

CHECKLIST OF CONSIDERATIONS AFFECTING SITE LAYOUT

Item	Consideration
Buildings	
❑ Offices for site manager	Avoid noise and dust
❑ Lock-up stores	Good view of site security
❑ Messroom, canteen	Clear of works
❑ Toilets, latrines	Downwind; Good drainage
❑ Gatekeeper/watchman	Good visibility
❑ Specialist shops (steel benders carpenters, blockmakers)	Near to their stores (or mixer) Inside crane radius
Plant and equipment	
❑ Cranes	Maximum weights to be lifted Load capacity at different radii; anchorages
❑ Hoists	Near to main work use
❑ Mixer	Near to aggregates and hoist Inside crane radius
❑ Generator	Isolated
❑ Blockmaker	Near mixer set-up
❑ Steel bender	Near steel store
❑ Power saws	Near timber store
Materials store	
❑ Cement	Under cover: near mixer
❑ Aggregates	Thief-proof: near mixer
❑ Timber	Inside crane radius
❑ Steel	Ease of delivery: inside crane radius
❑ Bricks and blocks	Ease of delivery: near the hoist
❑ Doors, windows, sanitary fittings, glass	Ease of delivery; safe storage area: remember - easily damaged
❑ Fuel	Isolated because of fire hazard
❑ Formwork	Room for fabrication, cleaning Inside crane radius
Access roads	
❑ Permanent access road and permanent parkings	Can they be used for site deliveries and unloading?
❑ Site entrance	Safety and traffic control
❑ Temporary site roads	Ground conditions; deliveries and unloading; parking areas
Hoarding and fencing	Safety of workers and public; security against theft

WHAT IS PRODUCTIVITY? 10

Getting the site layout right means that your workers can be productive. The next stage is to maximize site activity. The level of activity on site is a measure of how busy the workers are. This means finding out how much time is spent working and how much time is idle.

The lower the activity level on your site, the less money you will make on the contract, and if it is too low you might even make a loss. Workers who stand idle, either because they have nothing to do or because they are waiting for another operation to finish, still have to be paid. Moreover, the lower the general activity level on site, the longer the project will take. This also means that you lose money because the extra time you take to complete the project could have been spent starting up a new contract.

In order to improve your activity level on site, you need to be able to measure it. What you need is a true picture of what is happening, so that you can check up on the scope for improvement.

Productivity is a comparison between how much you have put into the project in terms of manpower, material, machinery or tools and the result you get out of the project. Productivity has to do with the <u>efficiency</u> of production. Making a site more productive means getting <u>more output</u> for <u>less cost</u> in <u>less time</u>.

Productivity covers every activity that goes into completing the construction site works, from the "planning" stage to the final site clearing. If the contractor can carry out these activities at <u>lower cost</u>, in <u>less time</u>, with <u>fewer workers</u>, or with <u>less equipment</u>, then productivity will be improved. There are many different ways of improving productivity on construction sites, fabrication yards and workshops. Some will offer a greater cost saving than others, but all should increase the final profit that a contractor can make on a contract. Your job as a manager is to look at the way each operation is being carried out, with a view to changing it to improve productivity.

REASONS FOR LOW PRODUCTIVITY

There may be several reasons why your activity count is low, and not all of them are the fault of your workers. It could be your own fault.

Here are some reasons for low activity:

- ❑ supervisors looking after too many people
- ❑ dissatisfied workers with a perceived grievance (for example, low pay)
- ❑ very heavy work on a hot day
- ❑ waiting for materials
- ❑ waiting for tools
- ❑ waiting for instructions
- ❑ machine breakdown
- ❑ waiting for another worker (or equipment operator) to finish so they can follow on (poor site layout)
- ❑ working in a confined space and getting in each other's way
- ❑ working gangs are out of balance (e.g. too many labourers to one mason)
- ❑ more people allocated to the task than needed.

INCREASING PRODUCTIVITY

Some factors which can lead to increased productivity are:

- ❑ efficient site layout
- ❑ more efficient tools
- ❑ incentive schemes
- ❑ more efficient use of equipment
- ❑ more efficient supervision
- ❑ use of more skilled workers: train your workforce
- ❑ reduced waiting time.

The following example shows different principles for raising productivity and some ways to achieve this goal.

Example: Three workers digging a narrow ditch, presently producing 8 m³ per day.

Ways of raising productivity	Examples
1. The same is produced, but less is put into production.	Due to the narrowness of the ditch, two workers can excavate as much per day as three do at present. By taking one worker out the cost is reduced. Production is the same so productivity is really raised.
2. More is produced with the same workforce.	The method of digging the ditch is improved so that they are now able to excavate 10 m³ instead of 8 m³ with the same tools and the same number of workers. The inputs are the same but production has gone up by 2 m³ so productivity is really raised.
3. More is produced and less is put into production.	One worker is released for other work and the remaining two excavate 10 m³ per day due to improved methods. Inputs are reduced by one worker, production is increased by 2 m³ so productivity is really raised.
4. More is produced *and* more is put into production, but the increase in production is bigger than the increase in inputs.	The workers' pay is increased by 20 per cent due to a bonus scheme, encouraging them to produce 40 per cent more. Since production has increased twice as much as the inputs, productivity is really raised.

When you are trying hard to make the site more efficient you must remember one thing:

The workforce cannot be expected to work at full pace all the time – <u>rests are necessary</u> to maintain efficient working.

Every site has an average level of activity, and every contractor should know what that level is when preparing the quotation and when calculating the allowables. As the owner or manager of your contracting business, you should break down the overall site plan into weekly plans, but the weekly target will only be achieved if daily targets are set and achieved. The site manager should ensure that daily targets are met by always planning the following day's work on the day before, using the allowables to calculate the average activity level.

At the end of the workday the site manager should compare the actual work done with the planned work target, in other words the "<u>actual</u> activity level" with the "<u>average</u> activity level". If the <u>actual</u> activity level is lower than the <u>average</u> activity level the job is <u>losing</u> money. The site manager should then look closer at the job, or activity, to find ways of improving productivity, and then replan the following day's work, incorporating the changes made to improve productivity on the jobs where money was being lost.

Now we will look at an example of how to use allowables when checking up on site activity.

The following cost allowables have been calculated for pouring concrete to foundations in Chapter 2. The unit is per cubic metre.

$$
\begin{array}{lll}
\text{Labour} & = & 7 \text{ NU} \\
\text{Plant} & = & 2 \text{ NU} \\
\hline
\text{Total} & = & 9 \text{ NU/m}^3
\end{array}
$$

The concrete gang consists of:

- 1 labourer loading aggregate and cement
- 1 labourer barrowing cement from shed and fetching water from a tank 20 m away
- 1 mixer operator
- 4 barrowing concrete (0.1 m^3 each barrow)
- 2 placing concrete
- 1 vibrator operator

The mixer is delivering 0.6 m^3 per mix.

The wage cost per day is:

8 labourers at 5 NU/day	=	40 NU
2 semi-skilled at 8 NU/day	=	16 NU

The plant cost per day is:

1 mixer at 10 NU/day	=	10 NU
1 vibrator at 2 NU/day	=	2 NU

Total daily cost of gang	=	68 NU (40+16+10+2)

In order to break even, the concrete gang must pour 7.5 m^3/day (68 NU divided by 9 NU/m^3 = 7.5 m^3).

If the gang cannot achieve this, the site manager must replan the job. It might be possible to cut down on the concrete gang and still achieve the same output.

Cement should already have been brought from the shed, and water should be made available at the mixer and added by the mixer operator (reduces with 1 labourer).

The mixer only turns out enough concrete to load 6 barrows fully. Therefore you could cut down on the labourers' barrowing to 3 (reduces by 1 labourer).

Could we do with only one man placing concrete? (reduces by 1 labourer).

Revised labour cost:

5 labourers at 5 NU/day	= 25 NU
2 semi-skilled at 8 NU/day	= 16 NU
Plant cost unchanged	= 12 NU

Revised daily cost of gang:	= 53 NU
To break even (53 divided by 9)	= 5.9 m^3

The concrete gang should be able to achieve 6 m^3/day (in fact, they should even be able to achieve the original 7.5 m^3/day).

The actual activity level is higher than the average activity level, so the job is making money. Productivity has been improved.

IMPROVING WORK METHODS   11

Improving productivity means changing the work methods so that the cost of carrying out an operation is reduced.

Improving the work method involves four steps:
1. <u>Select</u> the job/operation.
2. <u>Record</u> and describe the present method of doing the job.
3. <u>Improve</u> the method by thinking of better ways to do the job.
4. <u>Install</u> the new method on the job.

SELECT THE JOB/OPERATION

There may be several reasons for selecting an operation for improvement, such as:

high production costs

limiting factor for other activities

double handling of materials

not achieving the quality standards

danger and fatigue

low activity level.

Now we will look at these different reasons more carefully and see why you should select a certain operation for improvement.

High production costs

When it is clear that an operation becomes much more expensive than you had calculated.

The more an operation costs through wages, rental of equipment, material costs, etc., the less your profit. The extra cost may arise from more workers being needed than calculated, the operation taking longer than expected or the operation requiring more materials than estimated.

Limiting factor for other activities

When, for some reason, the operations are held up and there is a temporary delay.

While the project is held up you still have to pay the running costs, therefore any delay means using money that would otherwise have been profit. Examples are:

❑ lorries waiting to be loaded

❑ masons waiting for blocks

❑ the concreting gang waiting for steelfixers to finish.

Double handling of materials

When you find that materials for a particular operation are repeatedly wrongly located, so that they have to be picked up again and taken to their place of use.

This means extra costs in wages and additional delays.

The materials may also be damaged when handled: "move a pile of bricks three times and you'll have a pile of rubble".

Not achieving the quality standards

Your reputation as a competent contractor is at risk whenever the specified quality standards are not met.

Poor quality standards lead to low productivity because part of the work has to be done again or the contractor loses retention money (or substandard materials are returned to the manufacturer). Some of the reasons for not meeting the quality standards can be:

❑ wrong tools

❑ wrong materials

❑ wrong methods

❑ unclear instructions given.

Danger and fatigue

A building site, workshop or fabrication yard is a hazardous place, and when an operation is dangerous it means that people's lives are put at risk.

The risk can be reduced or eliminated by improving their method of work and providing proper equipment. One example is improved scaffolding. When the work becomes particularly strenuous, productivity also goes down because the workers operate at reduced capacity. People working under dangerous conditions tend to be very unproductive, worrying about their own safety.

A solution that is often used is to employ more workers for especially demanding operations. That is very useful when you have a special task demanding extra work but there is no acceptable substitute to making the workplace as safe as possible. For example, it may pay to add a few extra workers to the concreting gang at the time of a big pour, so that the work can be completed more quickly, but if productivity is low because your scaffolding is dangerous the solution is <u>not</u> to hire more workers but to make the scaffolding safer. You have a moral obligation to make the site as safe as possible and not to risk your workers' lives: in addition remember that safety also means higher profit through a more loyal and productive workforce.

Low activity level

If the activity level is generally too low on a site or on a certain job, action must be taken to raise it.

A low activity level means that a lot of money will be lost every day.

REMEMBER: Select only the operations which it will pay to improve. Changing a method or introducing a new one takes time and therefore costs money. There is no point in selecting an operation where the costs of improvement are greater than the savings for the improved method. Therefore, before choosing an operation for improvement make an assessment of:

how much it will cost in terms of money and time to change

how much saving will result from the improvement.

Compare the two and choose the operations that have a significantly higher potential for saving than the cost of introducing the new method.

RECORD AND DESCRIBE
THE PRESENT METHOD

The success of recording and describing the present method of doing the job depends on the accuracy with which facts are recorded, and the ease with which they can be studied. Notes should be kept simple and neat. The job studied should be broken down into its various operations.

For example, concreting could consist of the following operations:

❏ mixing

❏ transporting

❏ placing and finishing.

IMPROVE THE METHOD

Once the job has been recorded, it should be examined with the aim of devising an improved method. Depending on the cause (wrong location of materials, too few workers, poor supervision, etc.), an alternative, more cost-effective method is worked out. When working out a new method based on the recorded facts it is important to:

❏ consider as many alternatives as possible

❏ do proper calculations to be sure that the new method will be more cost effective.

INSTALL THE NEW METHOD

Timing
The correct time to install the new method will depend on the nature and complexity of the operation. For a new operation to be started on site, the timing will depend on the overall programme and progress of the construction project. If supervisors or workmen require training, some convenient period should be chosen to cause minimum interference with other related operations.

Pilot trials

It may be advisable to try out the new method on a small section of the works in order to:

- ❑ deal with unforeseen problems without disrupting the rest of the project
- ❑ test the effectiveness of operational and control procedures
- ❑ Train supervisors who will introduce the method elsewhere
- ❑ Convince the workforce of the merits of the method.

Check and evaluate

Once the new method is installed, it should be checked frequently and the results evaluated, in order to:

- ❑ identify and deal with unforeseen problems
- ❑ check any tendency to drift back to the old method
- ❑ identify further opportunities for improvement.

INCENTIVE SCHEMES 12

Methods of payment

The method used for paying the workers can have a major effect on productivity and efficiency on the site. If an effective payment scheme is introduced, profits can be increased through higher efficiency at the same time as the workers earn more.

There are many different ways of rewarding workers. Various methods of payment which reward workers with money, time off or both are:

❑ Daily (or hourly) wage – a fixed wage per day (or per hour)

❑ Piecework – a fixed amount per unit of work done

❑ Taskwork – finish the job and go home when ready

❑ Bonus schemes – extra reward for efficient work.

DAILY WAGE

The <u>advantage</u> of paying each worker a fixed daily rate is that it is easier to administer.

The <u>disadvantage</u> is that there are no extra incentives for the workers to improve their productivity.

PIECEWORK

The idea of this system is that the more the workers produce, the more they earn. Piecework requires considerable preparation, administration and supervision. The workers are paid according to the unit of work done, such as the number of blocks made, cubic metres excavated or square metres of roofing laid.

The unit rate will have been calculated for the cost allowables, but these may have to be altered slightly in practice.

The best policy is to work towards a set of rates which are regarded as fair both by workers and employer, and keep to them. Piecework can be used for gangs as well as for individuals.

TASKWORK

A lump sum is fixed for a complete task – for example fell a large tree, saw it into pieces and dispose of it. When it is completed the workers have earned their wage, and they may choose to:

move to another task and earn more, or

go home early and return next day.

The incentive is that the faster they get the job done the sooner they get paid, either in money or free time.

BONUS SCHEMES

A bonus is an extra reward for good performance which is added to the existing system of payment, whether it is daily wage, piecework or taskwork.

If you give bonuses to your workers in the right situations, there is a better chance that they will remain loyal to you and that they will try to work as efficiently as possible.

When should a bonus be given?

A bonus should be given when the result of the workers' extra effort gives the entrepreneur an advantage, for instance, getting the job done in time and thereby avoiding the need to pay liquidated damages. Try to avoid bonuses related to activities that are not critical, i.e. if early completion does not affect the overall completion date. The bonus should be given when the workers deserve it because they have done a good job, or when they need to be encouraged to carry out a difficult and demanding assignment. It is important to offer a bonus when it is genuinely earned and deserved.

A bonus at the wrong time or in the wrong situation is no incentive to better performance.

Examples of considerations are:

- [ ] Was the time schedule extremely tight?
- [ ] Was the ground very difficult to work in?
- [ ] Have the weather conditions been unusually bad?
- [ ] Is a subcontractor coming in on a certain date and does the contract stipulate that you have to pay if he cannot have access to the site area?
- [ ] Has the client been extremely difficult and caused the workers a lot of trouble?
- [ ] Has the public caused any problems for the workers when they have been performing their tasks?

What sort of bonus should be given?

There are many ways of giving your workers an extra reward. One possibility is a bonus system which is decided beforehand to encourage high efficiency. An example is to agree with the workers on a fixed bonus per item produced above the normal production rate, such as per brick laid above the normal number laid per day.

It is often better to give a team bonus rather than an individual bonus. Not only is it easier to calculate, but it will encourage a team of workers to help each other. The concrete placing gang will not achieve their targets if the mixer operators do not perform well. Earlier in this book, we saw an example of formwork carpenters being held up by the steelfixers. However, if you have several gangs working on the same sort of task, such as trenching and pipelaying, the competition between them is likely to lead to better performance.

A bonus does not need to be a way of encouraging workers beforehand. It can also be given to them after the task is completed to reward them for doing a good job. An example of this is when a contractor decides to give his people a little extra for being able to complete a job on time even though the weather was really bad. This sort of bonus is normally not planned but is decided by the contractor or the site agent when it is clear that a reward is appropriate.

What sort of bonus to give is left to the contractor's experience and negotiating skills. The most common way is to pay extra for a good job done or to give time off. However, there are other ways and the contractor (or manufacturer) who is a good personnel manager will know what is best in a given situation.

How much should be given?

This depends on how great the achievement is and on how often a bonus is given. If you have a policy of giving bonuses on a regular basis, the amount to be paid also has to fit in with this policy. Generally speaking, the more often a bonus is given, the smaller it has to be, otherwise you will have trouble in covering the costs.

How much to give is a very delicate question that needs careful consideration. If too much is given, the bonus may be considered too easy to get.

Another problem is that the same size of reward will be expected later. If the same bonus cannot be given later, this may have a negative effect on the morale of the workers.

On the other hand, if too little is given, the bonus is not considered a proper reward and it will not encourage the workers to increase their efficiency.

Introducing a bonus scheme –
points to remember

1. In most countries, there is a minimum daily wage. Whatever method of payment is used, the workers must be guaranteed the minimum wage. Also, the system of payment must not conflict with national labour regulations. It is important to check these before launching a system.

2. Most methods of payment encourage the workers to work more efficiently, i.e. to do more in less time. The result can be a decline in quality or the neglect of safety regulations. This must not be allowed. Therefore, when such a system is used, inspections of quality and safety may have to be made more often and more thoroughly.

3. The system should be fair. All the workers should have a reasonable chance of being rewarded for extra effort. A system which is not fair can be sabotaged by workers who feel that they lose out on it. In particular, don't forget your foremen and supervisors. They will probably have to work harder when a bonus scheme is in operation, and they should receive a suitable reward if it succeeds.

4. A good bonus system is easy to understand and the job is measured in an easy but objective way. The targets should be as simple as possible, for example you could, when

digging a trench, give the workforce a T-shaped template so that they can check for themselves if they have achieved the goal.

5. When a new system is introduced or changes are made to an existing one, it has to be explained properly to everybody involved. Then they know exactly what is expected from them. If a misunderstanding arises it can lead to conflict and spoil the system.

6. A system with a team bonus can be very efficient since it rewards the entire group. It encourages people to work together and it is often considered fair by the workers.

7. Different operations require different methods of payment. Factors to consider are:
   – Will it lead to savings?
   – Is the work easy to measure?
   – Can the system be easily administered?
   – Will the system be fair?

8. In some cases the workers are not able to reach their target because of delays caused by the management. In such cases the bonus should not be taken away from them, because the delay was beyond their control. This reinforces the need for good management if bonus systems are to be used. If the management is not efficient, frequent delays will occur and bonus money will have to be paid out without increased labour productivity.

9. Remember that a bonus system can cover good ideas as well as physical performance. Management is about getting things done with and through people. If your workers care about their skills, they are likely to have good ideas on how to improve the way things are done. You should not be afraid to discuss your project plans with them. If they come up with an idea that saves money, they will have earned a bonus and they will work harder to put the new idea into practice.

HEALTH AND SAFETY   13

High standards of safety should be an objective pursued in the same way and with as much vigour as other management objectives. Apart from the humanitarian aim of ensuring the well-being of all concerned, it is obvious that accidents and illness mean additional costs, and perhaps disruption of the contract.

Improving safety

The entrepreneur can assist in the prevention of accidents and thereby improve overall contract performance through the following methods:

1. Effective communication

   Communicating effectively with the workforce on accident prevention is often the key to a successful approach to safety improvement.

2. Record keeping

   Keeping records of the types of accident that occur most frequently, and why they occur, puts you in a better position to prevent them as you know what you are fighting against.

3. Motivation of the workforce

   Besides general steps such as providing information on accidents and their causes and consequences, some special motivation measures can be introduced, such as organizing a "safety" bonus for the workers or gangs with the best accident record.

4. Use of safety equipment

   Make sure that safety equipment is available when and where it is needed, insist that it is always used, and take disciplinary action against workers who refuse or frequently forget to use the equipment.

EXAMPLES OF ACCIDENTS

Accidents may, of course, occur in a number of ways. Some examples are:

> through the collapse of walls, parts of buildings (particularly during demolition), stacks of materials or stockpiles of excavated material

> through the collapse or overturning of ladders, scaffolds, stairs or beams

> by falls of objects, tools or pieces of work

> by falls of persons from ladders, stairs, roofs, scaffolds or buildings through hatches and windows or through other openings

> during loading, unloading, lifting, carrying and transporting loads

> on or in connection with vehicles of all kinds

> at power plant and power transmission machinery

> in the operation of railways

> on lifting appliances

> on welding and cutting equipment

> on compressed-air equipment

> by combustible, hot or corrosive materials

> by dangerous gases

> during blasting with explosives

> when using hand tools

> by stepping on sharp objects.

CAUSES OF ACCIDENTS

In the following list, the causes of accidents have been grouped together according to their nature:

Poor planning or organization

❏ Defects in technical planning
For example, a crane may be hired for a long period to carry out lifts up to 10 tonnes. If planning has failed to foresee that one lift of, say, 12 tonnes will be required it is a great temptation for a project manager to turn a blind eye while the crane driver switches off his governor and attempts to lift the load, rather than go to the additional expense of hiring a larger crane just for one lift

- ❑ Fixing unsuitable time-limits so that the workforce may have to work excessive overtime to keep to schedule. Tiredness is a major cause of accidents
- ❑ Assigning work to incompetent contractors
- ❑ Insufficient or defective supervision of the work
- ❑ Lack of cooperation between different trades:
  For example, if, due to poor planning, a plumber has to chase his pipework into a recently plastered wall the plasterer is going to get angry and the chances of a fight breaking out are high.

During the execution of the work

- ❑ Construction defects
- ❑ Use of unsuitable materials
- ❑ Defective processing of materials.

Equipment

- ❑ Lack of equipment
- ❑ Unsuitable equipment
- ❑ Defects in equipment
- ❑ Lack of safety devices or measures.

Management and conduct of the work

- ❑ Inadequate preparation of work
- ❑ Inadequate examination of equipment
- ❑ Imprecise or inadequate instructions from supervisors
- ❑ Unskilled or untrained operatives
- ❑ Inadequate supervision.

Workers' behaviour

- ❑ Irresponsible acts
- ❑ Unauthorized acts
- ❑ Carelessness.

QUALITY CONTROL   14

What has quality control to do with business management? Everything, if you want to stay in business more than a few weeks or months. Poor workmanship, materials or finishes are not going to be accepted by any sensible client, since the client is paying for the work to be completed to the standards laid down in the specification. Even if you get away with poor quality work on one job, people will notice and your reputation will suffer. When the building is complete and your client uses it daily, he will forget about the cheap price and the short construction period, and he will only remember the quality. In boom times almost any contractor can get work, but it is only the quality-conscious contractor who can keep a steady workload when conditions are more competitive.

Getting a reputation for quality does not need to cost you much money. It can even be free. The secret is to set up an efficient system of quality control which covers all your activities – on the site and in your yard or workshop. What really costs money is to have your work condemned by the client's representative. With a good quality control system, many potential problems can be sorted out before there is any question of repairing or replacing finished work.

The client's representative

Quality control on a site is usually enforced, ultimately, by the client's representative, for example a clerk of works. This person is responsible for seeing that the daily activities of the contractor result in an end product which satisfies the contract specifications. The clerk of works will ensure that the materials used for making concrete are up to standard, that they are mixed in the correct proportions, and that tests are made on samples of the mixed concrete.

Although it is one of the duties of the clerk of works to inspect the contractor's work to ensure the achievement of quality standards in accordance with specifications, it is the contractor who has the legal responsibility to deliver a quality product. What is more, it is the contractor who has to pay for demolishing and replacing faulty work.

So the job of your foreman or site supervisor is not to try to fool the clerk of works into accepting poor quality work. Their job is to make sure that your system of quality control is operated so efficiently that the clerk of works has no reason to complain. It is always cheaper to get quality right in the first place than have to spend a lot of time and money before getting a certificate of practical completion and the release of the retention money at the end of the project. This chapter gives advice on how to set up a quality control system for your business, and Chapter 14 of the workbook will help you to check on how effective your quality control system is in practice.

A QUALITY CONTROL SYSTEM

The purpose of setting up a quality control system is to make quality control a matter of routine. For example, all materials should be checked before they are offloaded at the site, yard or workshop. There should be a firm rule that substandard materials are never accepted, even if some kind of discount is offered. They should always be returned to, or collected by, the supplier. Your long-term reputation should be worth more to you than any small savings you could make by using substandard materials.

EYES AND EXPERIENCE

A good site supervisor will make full use of the two Es: eyes and experience, as front-line weapons in the battle for good quality. Some examples of the faults that you can pick up on a visual inspection of a construction site are:

❏ a very wet mix of concrete
❏ mixing of aggregates where piles of sand and stone overlap
❏ insufficient cover to reinforcing steel
❏ soil falling into foundations
❏ inadequate vibration of concrete.

If you employ subcontractors, instruct your site supervisor to keep a particularly close eye on their employees, because they do not have the same interest in the firm's long-term reputation as your own full-time staff. Some unsatisfactory work is difficult to spot until the job is complete.

Some examples of the points to check on finishing work are:

❏ roof flashings leaking
❏ plumbing joints leaking
❏ electrical connections badly made.

CONCRETE QUALITY CONTROL

Test cubes should be taken regularly. But small sites do not justify setting up a concrete laboratory, so they must be sent to a laboratory for testing. This takes time, and failed cubes mean that concrete already placed has to be demolished. This means you lose money in four ways:

❏ the cost of demolition

❏ wastage of original materials

❏ cost of rebuilding

❏ extra cost of speeding up other activities to make up for lost time.

Some contractors try to save money by cheating on the amount of cement in the mix. This is a dangerous practice. It usually shows up when the cubes are tested. More seriously, it may show up if the building fails and could result in a criminal charge against the contractor. Even if you order the full amount of cement for your project, you must still make sure that it actually goes into the concrete mix.

Simple tests

Some simple tests for site quality control are:

❏ The slump test: this tests the consistency of wet concrete by compacting it in a slump cone. If the specification states a 50 mm slump requirement, then any mix in excess of that should be rejected. This costs money, but is much cheaper than demolishing finished work. The main cause of failure in slump tests is the addition of too much water to make the mix easier to handle.

❑ Sand testing: the difference between sharp (or concreting) sand and fine (or plastering) sand can be told by feel. In order to find the silt content, a sample of sand can be placed in a bottle of water, shaken, and left to stand overnight. The silt will settle on top of the sand and a rough percentage can be calculated by observation.

❑ By breaking a few samples of blocks or bricks with a hammer, their strength and consistency can be estimated. If in doubt they should be sent to a laboratory for further testing.

Some work methods that can be tested and controlled on site are:

concrete mixing

concrete placing

steelfixing

shuttering-alignment and support

keeping excavations clean

sewer pipe alignment

correct placing of damp proof courses

proper compaction of fill

correct kerb alignment.

Some finishings that can be tested and controlled on site are:

the watertightness of roofs, doors, windows, gutters, plumbing joints

the application of the specified number of coats of paint

the smooth finish of joinery.

Conclusion

A NAME FOR QUALITY

Remember that there are three ways in which a client judges a contractor:

❑ cost (the amount of the bid)

❑ time (the period for completion)

❑ quality.

If you only compete on cost, you will always be trying to scrape a living and clients will keep trying to reduce your prices even when your bid is competitive. If you compete on time you will be in a better position to obtain work from clients for whom time really is equivalent to money, such as shop or factory owners. However, the best way to compete is to build a reputation for quality. Most clients will tell you that there are many contractors to choose from, and more new firms enter the market every day, but it is still difficult to find a contractor who can be trusted with a prestige job where quality is vital.

So we have deliberately ended this book with a chapter on quality because we want to leave you with a determination to make your firm the first choice for quality work in your area. If you manage to do this, you will find that your firm will still be winning contracts when your competitors complain that there is no work to be obtained even by bidding at a loss.

ILLUSTRATIONS OF PROPOSED
NEW BUILDINGS AT WARDOBOYO 3

Figure A: Site plan, block plan and key plan

Block plan 1:3,000 Key plan: 1:30,000

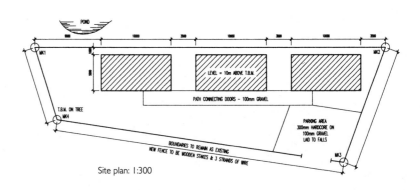

Site plan: 1:300

Figure B: General arrangement of building (scale 1:100)

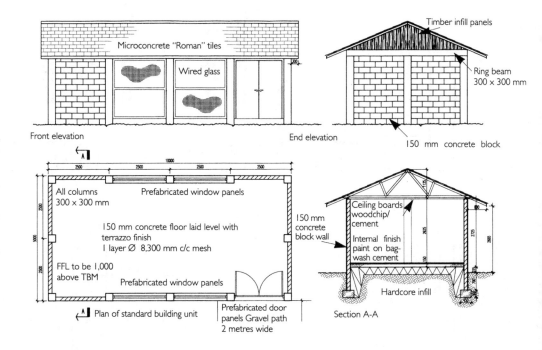

Microconcrete "Roman" tiles

Wired glass

Front elevation

Timber infill panels

Ring beam
300 × 300 mm

End elevation

150 mm concrete block

All columns
300 × 300 mm

Prefabricated window panels

150 mm concrete floor laid level with
terrazzo finish
1 layer Ø 8,300 mm c/c mesh

FFL to be 1,000
above TBM

Prefabricated window panels

150 mm
concrete
block wall

Ceiling boards
woodchip/
cement

Internal finish
paint on bag-
wash cement

Hardcore infill

Plan of standard building unit

Prefabricated door
panels Gravel path
2 metres wide

Section A-A

Figure C: Roof truss detail (scale 1: 40)

SCHEDULE OF TIMBER SIZES

A – D	42 × 112mm	1.5" × 4"
F – C	56 × 112mm	2" × 4"
C – E	56 × 112mm	2" × 4"
B – E	56 × 112mm	2" × 4"
Ridge Plate (C)	42 × 224mm	1.5" × 8"
Wall Plate (A)	42 × 112mm	1.5" × 4"
Fascia Board (F)	42 × 168mm	1.5" × 6"

* Roof trusses at 1 m centres
* 10 No. tile battens each side
28 × 28 mm, 1" × 1"

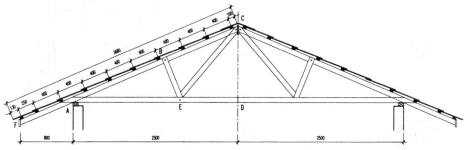

101

Figure D: Reinforced concrete details (scale 1:40)
(Schedule of bars not to scale)

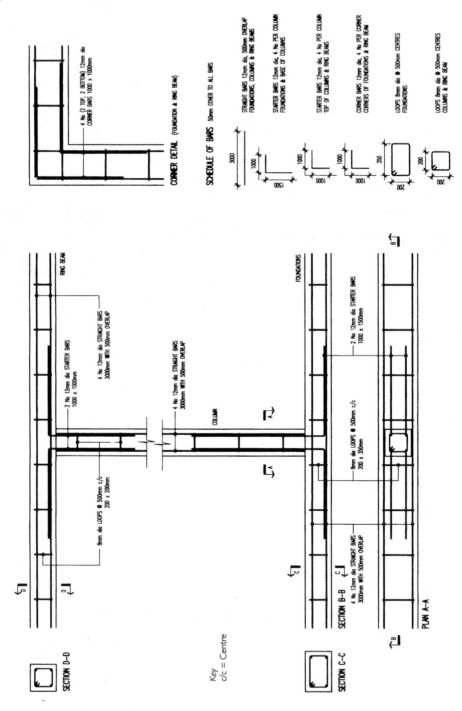